Quantum Computing

A Journey into the Next Frontier of Information and Communication Security

Edited by

Mohammad Hammoudeh, Abdullah T. Alessa,
Amro M. Sherbeeni, Clinton M. Firth, and
Abdullah S. Alessa

CRC Press
Taylor & Francis Group
Boca Raton London New York

CRC Press is an imprint of the
Taylor & Francis Group, an **informa** business

Designed cover image: Shutterstock

First edition published 2025
by CRC Press
2385 NW Executive Center Drive, Suite 320, Boca Raton FL 33431

and by CRC Press
4 Park Square, Milton Park, Abingdon, Oxon, OX14 4RN

CRC Press is an imprint of Taylor & Francis Group, LLC

ISBN: 978-1-032-75704-9 (hbk)
ISBN: 978-1-032-75705-6 (pbk)
ISBN: 978-1-003-47528-6 (ebk)

DOI: 10.1201/9781003475286

Typeset in Nimbus font
by KnowledgeWorks Global Ltd.

Contents

Quantum Computing

This book explores the exciting world of quantum computing, from its theoretical foundations to its practical applications, offering both non-technical and expert readers a comprehensive and accessible introduction to this cutting-edge technology that has the potential to revolutionize the way we process and transmit information.

Quantum Computing: A Journey into the Next Frontier of Information and Communication Security provides a comprehensive guide to the exciting and rapidly evolving field of quantum computing and communication security. The book starts by introducing the theoretical foundations of quantum mechanics and quantum computing, providing readers with a solid understanding of the principles behind this revolutionary technology. The book emphasizes the practical applications of quantum computing and its adoption strategies in response to the urgency of quantum readiness. While many books on the subject focus solely on the theory, this book explores the risks and opportunities of quantum computing, and how to prepare and adopt this technology. From there, the book explores various quantum concepts and their security applications, covering topics such as quantum-safe cryptography, standards, implications on artificial intelligence, and optimization.

The book is written for students, researchers, technology leaders, and professionals who work in the field of cybersecurity, communications, digital transformation, data analytics, and information systems. The book is suitable for researchers with various technical knowledge.

About the Editors

Mohammad Hammoudeh is the Saudi Aramco Cybersecurity Chair Professor at King Fahd University of Petroleum and Minerals. His research interest is in quantum communication technologies, quantum-safe encryption methods, and quantum adoption strategies in both IT and OT systems. He has hands-on experience in working with quantum hardware and algorithms. He developed practical ways of evaluating the real applications of new technology in the cyberspace such as quantum computing, digital twins, and blockchains.

Abdullah T. Alessa is a cybersecurity professional at Saudi Aramco cybersecurity operations. He has ten years of experience in various cybersecurity domains. He led multiple teams within the organization and currently serves as the head of Cryptography Solutions.

Amro M. Sherbeeni is working as a cybersecurity consultant in Saudi Aramco with over 25 years of experience in different IT & cybersecurity fields. He led multiple teams, projects, and functions within the organization and currently leading the Zero Trust adoption.

Clinton M. Firth is a partner at PwC focusing on digital, cyber and resilience. He is an experienced cybersecurity consulting executive, holding many senior and global roles for large consulting and system integrator firms. He has diverse experience working with commercial, government, and defense clients. Clinton leads a passionate team of industry specialists to address some of the most complex technology trust problems of our time.

Abdullah S. Alessa is an information technology and computing professional at Saudi Aramco computer operations. He has nine years of experience in several computing technology fields and lead various emerging technology projects in Saudi Aramco IT infrastructure.

List of Contributors

Bamidele Adebisi
Department of Engineering
Manchester Metropolitan University
United Kingdom

Abdullah T. Alessa
Information Protection Department
Saudi Aramco
Saudi Arabia

Abdulrahman AlRaimi
Computer Science and Engineering
    Department
Qatar University
Qatar

Saad Mohammed Anis
Computer Science and Engineering
    Department
Qatar University
Qatar

Madani Bezoui
CESI LINEACT
France

Ahcene Bounceur
Department of Information Systems
University of Sharjah
United Arab Emirates

Christoph Capellaro
EY Consulting GmbH
Germany

Sandrik Concepcion Das
Computer Science and Engineering
    Department
Qatar University
Qatar

Clinton M. Firth
PwC Middle East
United Arab Emirates

Mohammad Hammoudeh
Information and Computer Science
    Department
King Fahd University of Petroleum and
    Minerals
Saudi Arabia

Moumen Hamouma
LAMIE Laboratory Computer Science
    Department
University of Batna 2
Algeria

Ibrahim Kabir
Information and Computer Science
    Department
King Fahd University of Petroleum and
    Minerals
Saudi Arabia

Mostefa Kara
LIAP Laboratory University of Eloued
Algeria

Abdelkader Laouid
LIAP Laboratory University of Eloued
Algeria

Junseo Lee
Quantum AI Team Norma Inc.
South Korea

Foudil Mir
LIMED Laboratory University of Bejaia
Algeria

Amer Mosally
Information and Computer Science
  Department
King Fahd University of Petroleum and
  Minerals
Saudi Arabia

Harbaksh Singh
Architecture, Engineering & Emerging
  Technologies
Global Delivery Services Ernst & Young
India

Devrim Unal
KINDI Computing Research Center
Qatar University
Qatar

Danish Vasan
Information and Computer Science
  Department
King Fahd University of Petroleum and
  Minerals
Saudi Arabia

Preface

Welcome to an era supercharged by exponential technological advancement. As I write, I feel the powerful acceleration of Artificial Intelligence and Quantum Technologies, like engines revving as they gear up to shape our world. As a mathematician specializing in both AI and Quantum, I have the privilege of sitting in the driver's seat, innovating our approach to monumental challenges—such as cancer prevention, earthquake detection, space-time exploration, and the quest for general intelligence. I am particularly thrilled about Quantum for its potential to race past conventional boundaries, enabling us to tackle problems that are currently beyond our grasp. Quantum's impact is poised to speed through every layer of society, influencing strategic decisions and steering the course for corporations worldwide. As we navigate the twists and turns of integrating these powerful technologies, this textbook acts as an indispensable navigational tool, providing clear guidance on both the 'how' and the 'why' of these Quantum leaps.

At the dawn of the 21st century, the landscape of computational science and information security began a transformation under the shadow of quantum mechanics. As a result, quantum computing and quantum communications emerged in the evolution of these disciplines, promising to redefine our capabilities and secure our digital future. This publication comes at a critical juncture with the demand for a deeper understanding and practical knowledge of these groundbreaking technologies peaking. The intent herein is to provide a comprehensive resource charting the exciting confluence of quantum computing, quantum communications, and their impact across various business sectors. It is designed to bridge the gap between abstract quantum theories and their tangible applications, offering a detailed examination of how quantum principles are being applied to solve complex problems in everything from cryptography to computational biology. The text is engineered collaboratively by various industry experts to both inform and inspire, acting like a turbo boost for its readers on the art of the possible with quantum technologies. The material is structured to be both accessible and rigorous, ensuring that each chapter propels you further along this thrilling circuit.

It is my hope the pages within will expand your understanding and ignite a passion for what is unquestionably a pivotal moment of our era, the race of modern science and technology.

Dr. Kristin M. Gilkes
EY Global Quantum Leader
Associate Fellow – Artificial Intelligence, University of Oxford, Saïd Business School

1 A Primer on Quantum Computing and Quantum Communications

Christoph Capellaro

1.1 INTRODUCTION

Richard Feynman was an American physicist who participated in the famous Manhattan project that was initiated to build the first nuclear bomb in 1945. In a paper [1] he discussed the problems classical computers have with simulating physical effects, specifically when considering quantum physics. He stressed, that with classical computers and algorithms, it is possible to approach solutions of differential equations, but you would never be able to achieve exact results. To overcome this issue, he recommended to use a quantum computer that would use particles like photons or electrons and facilitate effects of quantum mechanics to manipulate those results, effectively performing some kind of computation. His paper even described the basic mathematics required for computation, but he wasn't sure whether technology could provide the necessary infrastructure required.

1.2 FIRST APPLICATIONS

The fact that quantum computers were only hypothetical at that point of time didn't refrain scientists from developing algorithms for such a machine. Interestingly, the first prominent results had all to do with information security. In 1984 the two IBM researchers Bennett and Brassard described the first protocol that could be used for distributing a shared secret between two distant parties [2]. The principle of quantum key distribution (QKD) was born. It is based on the effect of entanglement, which will be explained in more detail later. Consider two entangled photons. For any photons, you can measure their spin. Quantum physics ensures that if one of them is measured with an up-spin, the other one will be measured with a down-spin no matter how long the distance between these two photons is. It is important to notice that this effect cannot be used to transfer a message between two parties because nobody can determine which spin the first photon will take, but it's a perfect way to create a shared secret between two distant parties. To make this work, the two communicating parties need a quantum channel to exchange the entangled photons plus a control

DOI: 10.1201/9781003475286-1

channel. This control channel can be public but needs to provide authenticity of the sender, recipient, and messages.

The protocol according to Bennett and Brassard requires two different bases of measurement agreed upon between the two communicating parties, Alice and Bob. They measure their photons either in a basis consisting of a horizontal and a vertical axis or another one being X-shaped. To ensure that possible eavesdropping is detected, Alice documents what basis she was using for each of her qubits and then sends the entangled qubits to Bob. Bob confirms over the control channel that he received the qubits and measures them using arbitrary bases for each qubit. Then, Alice sends over the basis she selected for the measurement of each qubit. Bob returns those qubits for which he used the same basis as Alice. Now both have a shared secret, and due to the no-cloning theorem of quantum physics, nobody was able to take copy of the transmitted qubits.

There is still the chance that the attacker Eve measured the qubits that Alice sent to Bob during transmission. However, since Alice publishes her choice of basis only after Bob confirms the receipt, this could have happened only before Bob does his measurement. Hence there is a 50% chance that Eve's measurement would alter the qubits before Bob does his measurement since she used a different basis than Bob did. To detect this modification, Alice and Bob agree on an arbitrary set of control qubits and compare their results publicly. This gives them an idea, whether their channel had been intercepted.

This protocol of Bennett and Brassard is proven to be secure if its implementation is perfect. If Eve gets some side channel information about the axes used or the control bits chosen, there exist possible attacks to this scheme. Already in the early 90s of the 20th century, there had been successful implementations of this scheme using both fibre optics and over the air transmission.

In 1994 Peter Shor published his article about an algorithm that makes use of a hypothetical quantum computer to factorize large integers [3]. This was the first algorithm for a quantum computer that really struck the scientific community as it would have severe impact on information security if suitable hardware would be available. Clearly the RSA algorithm, whose security depends on the factorization of large integers would be affected, but with little modification also the discrete logarithm problem can be solved. This would basically affect almost all asymmetric cryptographic algorithms currently in use. So current asymmetric cryptography is at risk as soon as quantum hardware is strong enough. To make things worse, shortly after this Lov Grover published an algorithm on a quantum-based search in 1996, which can be used to attack symmetrical cryptography by accelerating brute force key search [8]. Thankfully this algorithm shows only a quadratic improvement compared to classical search algorithms. Hence squaring key lengths is sufficient to gain confidence in the strength of the used symmetric algorithms.

To complete the introduction of quantum algorithms we also need to mention algorithms that can be used to solve sparse linear equation systems [9] and quantum optimization algorithms. These algorithms open a wide field of applications in operation research and related areas, where optimization problems with various boundary conditions can be tackled. Several chapters of this book will show used cases in this area.

1.3 HOW DOES IT WORK?

These fascinating fields of application for quantum computers raise interest in understanding the difference between classical and quantum computing. The magic lies in qubits, superposition, entanglement, and interference, which we are going to explain in this section.

1.3.1 THE QUBIT

A key difference that is often mentioned when the principles of quantum computing are explained is that, that a qubit, which is the smallest unit of information a quantum computer can operate on, can represent not only one of the values 0 or 1, like a classical computer, but both values at the same time. But what does this mean?

To understand this, it is a good idea to introduce the Bloch sphere as a model to visualize a qubit. Felix Bloch was a physicist born in Switzerland in the early 20th century, who worked in the United States and briefly participated in the Manhattan project like Feynman did, but left it soon after, since he didn't like the military purpose. He later worked on magnetic resonance of atoms laying the basement for computer tomographs. In 1952, he was awarded the Nobel Prize for this work.

The Bloch sphere is basically a ball representing all possible states of a quantum system on its surface [5]. In praxis, the qubits of a quantum computer might also have a state that could be somewhere under the surface. This is due to errors, the quantum computer hardware engineers must cope with, but this shall not be part of our discussion here. So for us, the possible states of a qubit can be represented by the set of all points on the surface of the Bloch sphere. That's quite a number of different states for a single qubit to be in.

But why would we say that a qubit can be both 0 and 1 at a time? Here, another element of quantum mechanics comes into the game, which is called measurement. We can manipulate qubits (how to do this is explained in the following sections) to basically perform calculations, but if we want to read the result, we need to measure the state the qubits are in. Due to constraints in physics, measurement can be only in one direction. If we measure the spin of a quantum system such as a photon, it can be either up or down but nothing in between. Hence, as soon as we measure the state a qubit of our quantum computer is in, the quantum system decides to take one of two possible states.

In the representation of the Bloch sphere, measurement is usually taken along the z-axis passing through the north and south poles. By common agreement, we interpret these two poles of the Bloch sphere with the two states 0 and 1 of a classical bit, respectively. At the moment of measurement, the qubit decides for one of these two options. And the probability of being measured as either 0 or 1 depends on how close the actual state of the qubit is to either the north or the south pole. Is the distance of the actual state to the north pole quite tiny? If so, it is likely that we measure a 0, and vice versa. A state on the equator of our Bloch sphere results in a fifty-fifty chance of retrieving either a 0 or 1.

This is, where the expression that a qubit can be both 0 and 1 comes from. A more precise description of the situation would be to say that a qubit is in a state that can be measured with a certain probability p as 0 and with the probability $1 - p$ as 1. This gives us a first understanding, how a qubit looks like. In the next sections, we look at ways to manipulate qubits.

1.3.2 SUPERPOSITION

The fact that the state of a quantum system can be represented as a point on the surface of the Bloch sphere is also what characterizes superposition. Superposition here can be understood as the fact that a qubit can be measured with a certain probability of being 0 or 1 respectively. From the perspective of information theory, this state of a quantum system can be described as a linear combination of the vectors of the two basic states $|0\rangle$ and $|1\rangle$, representing the two directions from the center of the Bloch sphere to its north and south poles. The special notation of brackets shown here is called ket notation and goes back to Paul Dirac, a British mathematician and physician who worked on quantum physics in the early 20th century. This notation has a special purpose that we are not going to discuss further in the context of this book. What we would like to point out is that this notation is commonly used to nominate linear combinations of quantum states. Any point on the Bloch sphere can be interpreted as a vector Ψ, the linear combination of the vectors of the two basic states $|0\rangle$ and $|1\rangle$ in the form $|\Psi\rangle = \alpha|0\rangle + \beta|1\rangle$, with two complex numbers α and β and $|\alpha|^2 + |\beta|^2 = 1$. The probability of Ψ being measured as $|0\rangle$ is then given by $|\alpha|^2$ and the probability of Ψ being measured as $|1\rangle$ by $|\beta|^2$, respectively.

Superposition is often illustrated by the Schrödinger's cat experiment. Erwin Schrödinger, an Austrian physicist who won the Nobel Prize in Physics in 1933 alongside Paul Dirac for their contributions to quantum mechanics, devised this experiment to illustrates the situation that a single quantum element can be in an in-between state between two classically exclusive states. In his experiment, a cat is placed inside a sealed box along with a vial of cyanide, a hammer, a Geiger counter, a detector for α-particles and an instable atom that would emit an α-particle, with the probability of $1/2$. If the atom decays, the Geiger counter triggers the hammer to break the vial, releasing the cyanide and killing the cat. According to the principles of quantum mechanics, until the box is opened and observed, the cat exists in a superposition of being both alive and dead simultaneously. It is only when an observer opens the box, a process analogous to making a measurement in quantum physics, that the superposition collapses, and the cat is revealed to be either alive or dead. Before this observation, the cat remains in a quantum state that defies classical logic.

How does this relate to the model of the Bloch sphere introduced earlier in this section? Therefore, we add Cartesian coordinates to our model by putting the origin of coordinates into the center of the Bloch sphere, the z-axis will run along the direction of our vectors $|0\rangle$ and $|1\rangle$, $|0\rangle$ being in the direction of the positive z-axis and $|1\rangle$ in the direction of the negative z-axis. Any vector pointing to the equator of the Bloch sphere represents a superposition of $|0\rangle$ and $|1\rangle$. This is specifically true for the two vectors pointing in the direction of the positive and negative x-axis. These two vectors are designated with $|+\rangle$ and $|-\rangle$.

Usually, the qubits of a quantum computer are initialized in the state vector $|0\rangle$, and by the use of rotations we are able to transform this vector of the initial state to any other vector on the Bloch sphere.

Now after we learned what a single qubit looks like, we would like to understand, how two qubits can be brought in relation to each other.

1.3.3 ENTANGLEMENT

Superposition is a physical effect that distinguishes quantum computing from classical computation. Now, we want introduce a second element of quantum systems that is used as a building block for quantum computers – entanglement. Entanglement was first referred to by Albert Einstein and others in 1935 in a paper which had the intention to argue that quantum physics cannot be a proper description of reality. A. Einstein and his coauthors showed that a solution of Schrödinger's equations (which are basically used to describe quantum systems) allows an instant effect between two quantum systems that might even be separated by an arbitrary far distance. Instant in this context means that there is no time delay between the effect occurring in the one system and in the other, no matter how far the two systems are away from each other. Basically, this produces a kind of "faster-than-light" interaction. Einstein called this a "spooky action at a distance". This effect does not only exist (the Austrian physicist Anton Zeilinger was awarded a Nobel Prize in 2022 for his work about entanglement) but can be utilized in many ways, e.g., for quantum key distribution, which will be explained in a later chapter of this book.

To describe entanglement, the ket notation introduced above comes in handy. We can describe a quantum system consisting of the two qubits $|\Phi\rangle$ and $|\Psi\rangle$ with $|\Phi\Psi\rangle$. For example, two qubits representing the binary string 00 can be written in ket notation as $|00\rangle$ and two qubits representing the binary string 11 can be written as $|11\rangle$. By the way, we are using the big-endian notation of bitstrings throughout this section, meaning that the most significant bit of a bitstring is on its left. For example, the bitstring 10 represents the number 2. An example of an entangled state is then nominated with $\frac{1}{\sqrt{2}}(|00\rangle + |11\rangle)$. When we measure this quantum system, we will get the result 00 with a probability $\frac{1}{2}$ and the result 11 with the same probability. Results 01 or 10 have a probability of 0% in this case. Whenever the state of more than one qubit can only be expressed as a linear combination of states, we see entanglement. If their state can be described with a single vector, we have a separable state. For example, when the state $|10\rangle + |11\rangle = |1+\rangle$ is separable, the first qubit is in the state + and the second one in the state 1, whereas when the state $|00\rangle + |11\rangle$ is entangled, the states of the two qubits are interconnected.

1.3.4 INTERFERENCE

Interference is a property of physical systems that can be also examined on the level of quantum systems. A good interpretation of interference is given by waves. Waves of the same wavelength and amplitude can cancel each other out or they can reinforce each other depending on the phase. In the case that the phase of both waves is identical, they reinforce each other. If the phase is shifted by a half of the wavelength, the two waves cancel each other out.

The effect of interference is used in algorithms for quantum computers to increase the probability of desired results. To describe this, we interpret the set of possible states of a quantum computer with n qubits as an n-dimensional complex vector space, also called Hilbert space over complex numbers. We don't want to introduce the linear algebra here. In case you would like to read more about the Hilbert space, please be referred to the corresponding literature.

A state of the quantum computer is then a linear combination of vectors in this n-dimensional Hilbert space. What an algorithm for the quantum computer wants to achieve, is, to give the intended result of the calculation, which is represented as a specific vector in the n-dimensional Hilbert-space, a probability which is higher than other results. Interference is a possible mechanism that can be used to achieve this result.

Therefore, the programmer needs a universal set of gates that allow her or him to do the necessary operations. How this looks like for a quantum computer is explained in the next section.

1.4   A UNIVERSAL SET OF QUANTUM GATES

After we learned about the quantum physics that is behind qubits, we would like to understand, how we can perform calculations. We are looking for a universal set of logical gates that enable the implementation of quantum algorithms. We learned already in the last section that the state, a qubit can be in, is best described with a vector in the Hilbert space. Hence, we represent the two states of the qubit that refer to the classical states 0 and 1 as $|0\rangle = \begin{pmatrix} 1 \\ 0 \end{pmatrix}$ and $|1\rangle = \begin{pmatrix} 0 \\ 1 \end{pmatrix}$. Operations on a single qubit are then performed using unitary matrices of complex numbers. Unitary means that the inverse of a matrix is equivalent to its complex conjugate. Using unitary matrices for operations on state vectors ensures that no matter what operation we are performing, the vector will always direct to a point on the Bloch sphere.

The simplest operation is the "do nothing" operation. It is represented by the identity matrix, e.g.: $I|0\rangle = |0\rangle \Leftrightarrow \begin{pmatrix} 1 & 0 \\ 0 & 1 \end{pmatrix} \cdot \begin{pmatrix} 1 \\ 0 \end{pmatrix} = \begin{pmatrix} 1 \\ 0 \end{pmatrix}$. If we want to flip the states $|0\rangle$ and $|1\rangle$, which actually results in a NOT gate, we use the X matrix $X = \begin{pmatrix} 0 & 1 \\ 1 & 0 \end{pmatrix}$, $X|0\rangle = |1\rangle$ and $X|1\rangle = |0\rangle$. From a geometrical point of view, the X matrix performs a rotation of a vector on the Bloch sphere around the x-axis about an angle of 180°. Qubits allow any kind of rotations around axes passing the center of the Bloch sphere. Another example is the phase operation $P_\theta = \begin{pmatrix} 1 & 0 \\ 0 & e^{i\theta} \end{pmatrix}$, which rotates a vector around the z-axis about the angle θ. In case of $\theta = 180°$ we get $e^{i\pi} = -1$, which results in a rotation around the z-axis about 180°. This is also called the Z gate in quantum computing.

To achieve superposition the Hadamard operation is used. Its matrix is defined as $H = \frac{1}{\sqrt{2}} \begin{pmatrix} 1 & 1 \\ 1 & -1 \end{pmatrix}$. Some simple calculation shows that $H|0\rangle = |+\rangle$:

$$\frac{1}{\sqrt{2}} \begin{pmatrix} 1 & 1 \\ 1 & -1 \end{pmatrix} \begin{pmatrix} 1 \\ 0 \end{pmatrix} = \frac{1}{\sqrt{2}} \begin{pmatrix} 1 \\ 1 \end{pmatrix}$$

In the representation of the Bloch sphere that we introduced above, this vector points into the direction of the positive x-axis, hence it is the vector $|+\rangle$. The same calculation in ket notation looks like this: $\frac{1}{\sqrt{2}}(|0\rangle + |1\rangle) = |+\rangle$. You may convince yourself that $H|+\rangle = |0\rangle$, $H|1\rangle = |-\rangle$, and $H|-\rangle = |1\rangle$.

In order to work with numbers that are larger than one, we need more than one qubit and an operation between two different qubits. The elementary logical gate to do so is the controlled-not or CNOT gate, which in case of quantum computing can be achieved by a controlled flip of a qubit. To describe this mathematically, we use a two-dimensional Hilbert space. A state consisting of two qubits is then represented by a four-dimensional vector. Consider, e.g., our two qubits are in the state

$|0\rangle \otimes |1\rangle = |01\rangle$, we get the vector $\begin{pmatrix} 0 \\ 1 \\ 0 \\ 0 \end{pmatrix}$ in the Hilbert space. Here the operator $\otimes$

stands for the tensor product. It is defined as follows: $\begin{pmatrix} b_1 \\ b_2 \end{pmatrix} \otimes \begin{pmatrix} a_1 \\ a_2 \end{pmatrix} = \begin{pmatrix} b_1 \cdot a_1 \\ b_1 \cdot a_2 \\ b_2 \cdot a_1 \\ b_2 \cdot a_2 \end{pmatrix}$.

Operations in the two-dimensional Hilbert space are performed by (4×4) matrices. The matrix for our CNOT gate looks as follows: $\begin{pmatrix} 1 & 0 & 0 & 0 \\ 0 & 0 & 0 & 1 \\ 0 & 0 & 1 & 0 \\ 0 & 1 & 0 & 0 \end{pmatrix}$. We leave it to you, to convince yourself that this keeps the first qubit unchanged and flips the second qubit, if the first qubit equals $|1\rangle$.

An interesting effect is observed when the control qubit of a CNOT gate is in superposition. Let's consider $CNOT(|1+\rangle) = CNOT(|1\rangle \otimes (|0\rangle + |1\rangle)) = CNOT(|10\rangle + |11\rangle) = CNOT(|10\rangle) + CNOT(|11\rangle) = |10\rangle + |01\rangle$. Now the two qubits are entangled! The mechanism behind this effect is called phase kickback, which we are not going to discuss in this section. Please be referred to standard literature to read more about this topic [7]. Below we will show, how this effect can be used for our advantage.

With I, X, H, Z, P_θ, and CNOT gates, we have a universal set of quantum gates for a quantum computer. In praxis, quantum computers offer a larger set of gates to the programmers' convenience, but for the purpose of this introduction we stick to the above.

1.5 HOW CAN QUANTUM COMPUTING BE FASTER THAN CLASSICAL COMPUTING?

The effects of quantum physics can be used to accelerate the computation of a solution for some specific problems. The time advantage materializes, when we would have to do the same calculation several times in classical computing, in contrast to perform this calculation only once for quantum registers that are in superposition. An easy way to showcase how this works is Deutsch's algorithm [6].

Consider a function from one bit to one bit, $f : \{0,1\} \to \{0,1\}$. This function can be either linear or constant. In the literature, the term "balanced" is often used instead of "linear". There are four different possibilities:

$$f_1(0) = f_1(1) = 0$$
$$f_2(0) = 0 \text{ and } f_2(1) = 1$$
$$f_3(0) = 1 \text{ and } f_3(1) = 0$$
$$f_4(0) = 1 \text{ and } f_4(1) = 1$$

where the functions f_1 and f_4 are constant, whereas the functions f_2 and f_3 are linear. To determine on a classical computer, whether $f(x)$ is linear of constant, we compute f for two different values of x and analyze the outcome. Now let's see, what we can do with a quantum computer.

We implement a quantum oracle that predicts the behavior of function f with one function call of $f(x)$. This quantum oracle maps an initial state $|x\rangle \otimes |y\rangle$ to $|x\rangle \otimes |y\rangle \oplus f(x)\rangle$. Here the operator $\oplus$ stands for the addition modulo 2 which is equivalent to the binary operator XOR.

Now, let's go through the steps of the algorithm:

1. We are preparing the quantum computer with the initial state $|y\rangle \otimes |x\rangle = |-\rangle \otimes |+\rangle = (|0\rangle - |1\rangle) \otimes (|0\rangle + |1\rangle)$.
2. Now we map $|y\rangle$ to $|y\rangle \oplus f(x)\rangle$. Depending on the chosen f, there are four options:

$$f = f_1 : |y\rangle \otimes |x\rangle \to (|y\rangle \oplus |0\rangle) \otimes |x\rangle = |y\rangle \otimes |x\rangle = (|0\rangle - |1\rangle) \otimes (|0\rangle + |1\rangle)$$
$$f = f_2 : |y\rangle \otimes |x\rangle \to CNOT(|y\rangle \otimes |x\rangle) = CNOT(((|0\rangle - |1\rangle) \otimes (|0\rangle + |1\rangle)))$$
$$= CNOT(|00\rangle + |01\rangle - |10\rangle - |11\rangle) = |00\rangle + |11\rangle - |10\rangle - |01\rangle$$
$$= (|0\rangle - |1\rangle) \otimes (|0\rangle - |1\rangle)$$
$$f = f_3 : |y\rangle \otimes |x\rangle \to CNOT(X(|y\rangle) \otimes |x\rangle) = CNOT(((|1\rangle - |0\rangle) \otimes (|0\rangle + |1\rangle)))$$
$$= CNOT(|10\rangle + |11\rangle - |00\rangle - |01\rangle) = |10\rangle + |01\rangle - |00\rangle - |11\rangle$$
$$= -(|0\rangle - |1\rangle) \otimes (|0\rangle - |1\rangle)$$
$$f = f_4 : |y\rangle \otimes |x\rangle \to (|y\rangle \oplus |1\rangle) \otimes |x\rangle = |y\rangle \otimes |x\rangle = (|0\rangle - |1\rangle) \otimes (|0\rangle + |1\rangle)$$

3. Thanks to the phase kickback, the qubit x now contains all the information we need to decide on the nature of function f. To get the result, we need to

transform the qubit x back to a basis state by applying a Hadamard gate: $x \rightarrow H(x)$. Let's have a look at the possible results:

$$f = f_1 : H(x) = H(|0\rangle + |1\rangle) = |0\rangle$$
$$f = f_2 : H(x) = H(|0\rangle - |1\rangle) = |1\rangle$$
$$f = f_3 : H(x) = H(|0\rangle - |1\rangle) = |1\rangle$$
$$f = f_4 : H(x) = H(|0\rangle + |1\rangle) = |0\rangle$$

4. The last step in the algorithm is to measure qubit x. When it's 0, we know that our function f was constant; when it's 1, we know it was linear.

1.6 CONCLUSION

There is a certain misconception that quantum computers would be able to solve problems that classical computers can't. From a theoretical perspective this is not true. All the calculations that we introduced above can be simulated on classical computer. It might take much longer to do so on a classical computer, but from a theoretical point of view, it is still solvable. Certainly, there is a difference in the complexity that some problems have for quantum computers versus classical computing.

As with other questions in complexity theory, quantum computational complexity is not yet fully understood. We would like to make some key remarks in this context. First of all, quantum computing is probabilistic. You can't look into the computational process, in order to get a result, you need to measure. At this point, all superpositions collapse and you receive one of possibly many results, each of them having a certain probability. The art of quantum computer programming hence is to give the intended results a higher probability. This makes quantum computers specifically useful to solve problems, where it is difficult to find a solution but is easy to verify that a solution is correct as soon as you got one. The factorization of large integers is a good example.

According to our current understanding, in complexity theory it is assumed that NP (NP is a complexity class that represents the set of all decision problems for which the instances where the answer is "yes" have proofs that can be verified in polynomial time) is included in BQP (this is the set of problems that can be solved in bounded-error polynomial time using quantum computing). In other words, BQP includes problems that can be solved by a quantum computer in polynomial time with limited error probability. This gives quantum computing great prospects, as most optimization problems are considered to be NP. Examples are the travelling salesman problem, the knapsack problem, etc. For further reading you may refer to Ref. [4].

REFERENCES

1. Richard P. Feynman, "Simulating physics with computers". International Journal of Theoretical Physics 21, nos. 6–7 (1982): 467–488.

2. C. H. Bennett and G. Brassard, "Quantum cryptography: Public key distribution and coin tossing". Proceedings of IEEE International Conference on Computers, Systems and Signal Processing, New York, 175, (1984): p. 8.

3. P. W. Shor, "Algorithms for quantum computation: Discrete logarithms and Factoring". Proceedings 35th Annual Symposium on Foundations of Computer Science. IEEE Computer Society Press (1994): pp. 124–134.

4. https://cs.uwaterloo.ca/~watrous/Papers/QuantumComputationalComplexity.pdf

5. Felix Bloch, "Nuclear induction". Physical Review 70, nos. 7–8 (Oct 1946): 460–474.

6. David Deutsch, "The Church-Turing principle and the universal quantum computer". In Proceedings of the Royal Society of London (1985): p. 97.

7. https://learning.quantum.ibm.com/course/fundamentals-of-quantum-algorithms/phase-estimation-and-factoring

8. Lov K. Grover, "A fast quantum mechanical algorithm for database search". Proceedings of the Twenty-Eighth Annual ACM Symposium on Theory of Computing - STOC '96. Philadelphia, PA: Association for Computing Machinery (1996): pp. 212–219.

9. Aram W. Harrow, Avinatan Hassidim, and Seth Lloyd, "Quantum algorithm for linear systems of equations". Physical Review Letters 103, no. 15 (2008).

2 Basics of Quantum Computing

Ahcene Bounceur, Mohammad Hammoudeh,
Bamidele Adebisi, and Moumen Hamouma

2.1 INTRODUCTION

Over the past few years, the technology sector has experienced a significant transformation driven by the rapidly growing discipline of quantum mechanics. Quantum technology, with its potential for exceptional processing capacity and unmatched accuracy, is transforming various industries, including encryption and drug development. This chapter introduces the foundational principles and essential concepts that underpin quantum computing, equipping the readers with the knowledge to navigate this field with clarity.

In 1998, Isaac Chuang of the Los Alamos National Laboratory, Neil Gershenfeld of the Massachusetts Institute of Technology (MIT), and Mark Kubinec of the University of California at Berkeley created the first quantum computer (2-qubit) that could be loaded with data and output a solution [1]. In December 2023, physicists, for the first time, reported the entanglement of individual molecules, which may have significant applications in quantum computing. Also in December 2023, scientists at Harvard successfully created "quantum circuits" that correct errors more efficiently than alternative methods, which may potentially remove a major obstacle to practical quantum computers.

Quantum programming originates from the fundamental unit of matter, the atom. Figure 2.1 illustrates the concept of an atom containing an electron. An electron has the ability to exist in two distinct orbits. Two states, state 0 and state 1, can be defined based on the occupied orbit. Hence, this technique can be utilised for encoding binary data.

To make the electron change its orbit, one must send a signal with a period T (Figure 2.1(1)). This is done in reality by using a laser. And to bring it back to its initial orbit, it is sufficient to send the inverse of the signal with the same period (Figure 2.1(2)). There is another case where we can send a signal with a half period $(T/2)$. This leads to a situation where the electron occupies the first orbit and the second orbit alternatively. This operation is very fast and exceeds the speed of light to the point where it is considered that the electron occupies both orbits at the same time. This situation is called **Superposition** which is presented in Section 2.7.1.

DOI: 10.1201/9781003475286-2

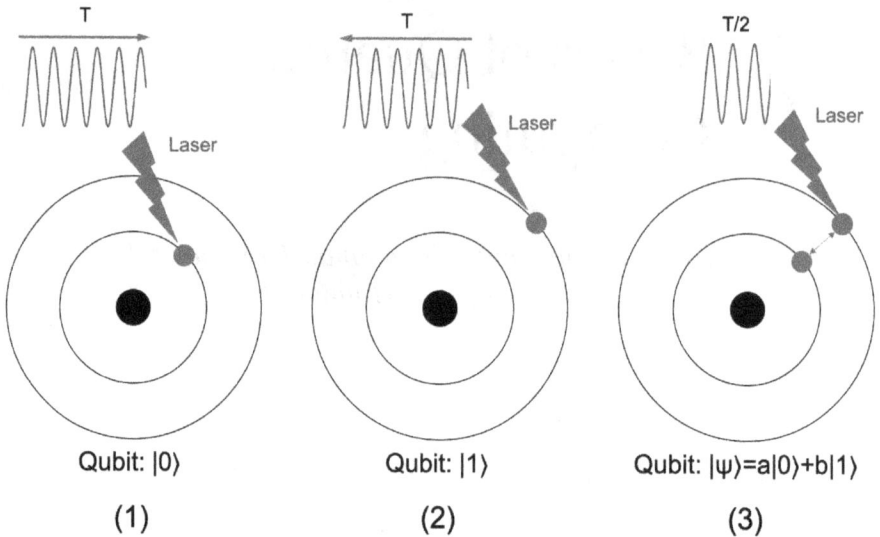

T

Laser

Qubit: |0⟩

(1)

T

Laser

Qubit: |1⟩

(2)

T/2

Laser

Qubit: |ψ⟩=a|0⟩+b|1⟩

(3)

Figure 2.1 Atom and qubit.

Superposition is crucial in quantum programming as it allows to design algorithms running very fast and with a much lower complexity compared with the one of classical algorithms. With a period of $T/2$, it is said that the electron has a probability of 50% to occupy the first orbit and a probability of 50% to occupy the second orbit. It is possible to change these probabilities, for example, 30% the electron occupies the first orbit and 70% the electron occupies the second orbit. This can be done by changing the period of the signal.

The qubit is the quantum equivalent of the classical bit. Qubits, unlike classical bits, have the extraordinary capability to exist in superposition states of 0 and 1. This allows them to store and manipulate large amounts of information simultaneously. To provide you with a comprehensive understanding of qubits, this chapter presents the Bloch sphere, a powerful geometric model that clarifies the behaviour of qubits in three-dimensional space [2–6].

The chapter also investigates the intricate world of quantum gates, which are the fundamental building blocks of quantum algorithms. Just as classical logic gates manipulate bits to perform operations, quantum gates manipulate qubits to execute quantum operations such as rotations, flips, and entanglements. By examining these quantum gates, the reader will gain an understanding of their role in quantum computation and their transformative potential for solving complex computational problems.

This chapter not only explains the basics of single-qubit and two-qubit systems but also explores generalised cases with n number of qubits. The power and versatility of quantum computation in action are illustrated through examples and practical demonstrations. Additionally, the chapter introduces the concept of quantum functions and oracles, essential components in quantum algorithms, enabling the manipulation and transformation of quantum states to achieve desired computational outcomes.

2.2 A QUBIT

The basic unit of information in a classical computer is the bit, which is represented either by 0 or 1. By combining multiple bits, one can encode integers; for instance, the binary representation of 25 is 11001. Similarly, in quantum computing, we begin with two fundamental quantum states: $|0\rangle$ and $|1\rangle$. These states represent the following vectors:

$$|0\rangle = \begin{pmatrix} 1 \\ 0 \end{pmatrix}$$

and

$$|1\rangle = \begin{pmatrix} 0 \\ 1 \end{pmatrix}$$

Also, the qubits $|0\rangle$ and $|1\rangle$ can be written as follows:

$$\begin{aligned}
|0\rangle &= 1|0\rangle + 0|1\rangle \\
|0\rangle &= (1+0i)|0\rangle + (0+0i)|1\rangle \\
|1\rangle &= 0|0\rangle + 1|1\rangle \\
|1\rangle &= (0+0i)|0\rangle + (1+0i)|1\rangle
\end{aligned} \tag{2.1}$$

What is new and fundamental is that these two states $|0\rangle$ and $|1\rangle$ can be superimposed. A qubit is a quantum state obtained through a linear combination:

$$|\psi\rangle = \alpha|0\rangle + \beta|1\rangle$$

Therefore,

$$|\psi\rangle = \alpha|0\rangle + \beta|1\rangle = \alpha\begin{pmatrix} 1 \\ 0 \end{pmatrix} + \beta\begin{pmatrix} 0 \\ 1 \end{pmatrix} = \begin{pmatrix} \alpha \\ \beta \end{pmatrix}$$

where α and β are complex numbers $\alpha \in \mathbb{C}$ and $\beta \in \mathbb{C}$.

Consider states $|\psi\rangle = \alpha|0\rangle + \beta|1\rangle$, where the norm is equal to 1. This means:

$$|\alpha|^2 + |\beta|^2 = 1$$

If $z = a + ib$ is a complex number (with $a, b \in \mathbb{R}$), then its norm $|z|$ is the positive real number defined by $|z|^2 = a^2 + b^2$.

One of the fundamental aspects of quantum physics is that the coefficients α and β of the quantum state $|\psi\rangle = \alpha|0\rangle + \beta|1\rangle$ cannot be measured directly.

Consider a normalised quantum state. The measurement of the quantum state $|\psi\rangle$ yields one of the classical bits, 0 or 1:

$$\begin{cases} 0 & \text{with a probability } |\alpha|^2 \\ 1 & \text{with a probability } |\beta|^2 \end{cases}$$

where, $|\alpha|^2 + |\beta|^2 = 1$.

As an example,

$$|\psi\rangle = \frac{1}{\sqrt{2}}|0\rangle + \frac{1}{\sqrt{2}}|1\rangle$$

is a qubit with the norm 1:

$$|\alpha|^2 + |\beta|^2 = \left|\frac{1}{\sqrt{2}}\right|^2 + \left|\frac{1}{\sqrt{2}}\right|^2 = \frac{1}{2} + \frac{1}{2} = 1$$

and the measurement of the quantum state $|\psi\rangle$ yields one of the classical bits, 0 or 1:

$$\begin{cases} 0 & \text{with a probability } 1/2 = 0.5 \\ 1 & \text{with a probability } 1/2 = 0.5 \end{cases}$$

Note that the measurement of a quantum state $|\psi\rangle$ irreversibly disturbs it. This is a physical phenomenon known as *Wave Function Collapse*. If the measurement resulted in the bit 0, then the state $|\psi\rangle$ has collapsed to $|0\rangle$; if the measurement resulted in the bit 1, then $|\psi\rangle$ has collapsed to $|1\rangle$. In other words, once measured, a qubit loses much of its utility.

In summary, starting from a state $|\psi\rangle = \alpha|0\rangle + \beta|1\rangle$ with $\alpha, \beta \in \mathbb{C}$ such that $|\alpha|^2 + |\beta|^2 = 1$:

- The coefficients α and β cannot be directly measured.
- Measurement of ψ yields either 0 with a probability of $|\alpha|^2$ or 1 with a probability of $|\beta|^2$.
- The measurement transforms the qubit $|\psi|$ into either $|0\rangle$ or $|1\rangle$, and the coefficients α and β vanish after measurement.

2.3 GATES

A quantum computer generates qubits and applies transformations to them, referred to as *gates* in a circuit. We begin by transforming a single qubit.

2.3.1 GATE X (NOT GATE)

The X gate, also known as the NOT gate, is a transformation that swaps the two basic quantum states:

$$|0\rangle \xmapsto{X} |1\rangle \quad \text{and} \quad |1\rangle \xmapsto{X} |0\rangle$$

Moreover, this transformation is linear, causing the X gate to exchange the two coefficients of a quantum state:

$$|\psi\rangle = \alpha|0\rangle + \beta|1\rangle \xrightarrow{\;X\;} \beta|0\rangle + \alpha|1\rangle$$

In terms of vectors, this transformation is expressed as:

$$\begin{pmatrix} 1 \\ 0 \end{pmatrix} \xrightarrow{\;X\;} \begin{pmatrix} 0 \\ 1 \end{pmatrix} \qquad \begin{pmatrix} 0 \\ 1 \end{pmatrix} \xrightarrow{\;X\;} \begin{pmatrix} 1 \\ 0 \end{pmatrix} \qquad \begin{pmatrix} \alpha \\ \beta \end{pmatrix} \xrightarrow{\;X\;} \begin{pmatrix} \beta \\ \alpha \end{pmatrix}$$

The matrix representation of the X gate is therefore:

$$X = \begin{pmatrix} 0 & 1 \\ 1 & 0 \end{pmatrix}$$

since,

$$\begin{pmatrix} 0 & 1 \\ 1 & 0 \end{pmatrix} \begin{pmatrix} \alpha \\ \beta \end{pmatrix} = \begin{pmatrix} \beta \\ \alpha \end{pmatrix}$$

The different outputs for different input states are presented in Section 2.3.4.1.

2.3.2   GATE H (HADAMARD GATE)

The Hadamard gate is denoted by H and the following circuit:

$$-\boxed{H}-$$

The H gate is the linear transformation defined by:

$$|0\rangle \xrightarrow{\;H\;} \frac{1}{\sqrt{2}}|0\rangle + \frac{1}{\sqrt{2}}|1\rangle \quad \text{and} \quad |1\rangle \xrightarrow{\;H\;} \frac{1}{\sqrt{2}}|0\rangle - \frac{1}{\sqrt{2}}|1\rangle$$

We can write,

$$|0\rangle \xrightarrow{\;H\;} \frac{1}{\sqrt{2}}(|0\rangle + |1\rangle) \quad \text{and} \quad |1\rangle \xrightarrow{\;H\;} \frac{1}{\sqrt{2}}(|0\rangle - |1\rangle)$$

For $|\psi\rangle = \alpha|0\rangle + \beta|1\rangle$, the $H(|\psi\rangle)$ is given as follows:

$$\begin{aligned} H(|\psi\rangle) &= \alpha\left(\frac{1}{\sqrt{2}}|0\rangle + \frac{1}{\sqrt{2}}|1\rangle\right) + \beta\left(\frac{1}{\sqrt{2}}|0\rangle - \frac{1}{\sqrt{2}}|1\rangle\right) \\ &= \frac{\alpha}{\sqrt{2}}|0\rangle + \frac{\alpha}{\sqrt{2}}|1\rangle + \frac{\beta}{\sqrt{2}}|0\rangle - \frac{\beta}{\sqrt{2}}|1\rangle \end{aligned} \tag{2.2}$$

Then,

$$H(|\psi\rangle) = \frac{\alpha + \beta}{\sqrt{2}}|0\rangle + \frac{\alpha - \beta}{\sqrt{2}}|1\rangle$$

In terms of vectors, this transformation is written as:

$$\begin{pmatrix} 1 \\ 0 \end{pmatrix} \xrightarrow{\;H\;} \frac{1}{\sqrt{2}} \begin{pmatrix} 1 \\ 1 \end{pmatrix} \qquad \begin{pmatrix} 0 \\ 1 \end{pmatrix} \xrightarrow{\;H\;} \frac{1}{\sqrt{2}} \begin{pmatrix} 1 \\ -1 \end{pmatrix} \qquad \begin{pmatrix} \alpha \\ \beta \end{pmatrix} \xrightarrow{\;H\;} \frac{1}{\sqrt{2}} \begin{pmatrix} \alpha + \beta \\ \alpha - \beta \end{pmatrix}$$

The matrix representation of the H gate is therefore,

$$H = \frac{1}{\sqrt{2}} \begin{pmatrix} 1 & 1 \\ 1 & -1 \end{pmatrix}$$

since,

$$\frac{1}{\sqrt{2}} \begin{pmatrix} 1 & 1 \\ 1 & -1 \end{pmatrix} \begin{pmatrix} \alpha \\ \beta \end{pmatrix} = \frac{1}{\sqrt{2}} \begin{pmatrix} \alpha + \beta \\ \alpha - \beta \end{pmatrix}$$

The notations $|+\rangle$ and $|-\rangle$ are used to represent the following states:

$$|+\rangle = \frac{1}{\sqrt{2}}(|0\rangle + |1\rangle) \qquad |-\rangle = \frac{1}{\sqrt{2}}(|0\rangle - |1\rangle)$$

then we conclude that,

$$|0\rangle \xrightarrow{\;H\;} |+\rangle \qquad |1\rangle \xrightarrow{\;H\;} |-\rangle$$

In terms of vectors:

$$|0\rangle = \begin{pmatrix} 1 \\ 0 \end{pmatrix} \xrightarrow{\;H\;} \frac{1}{\sqrt{2}} \begin{pmatrix} 1 & 1 \\ 1 & -1 \end{pmatrix} \begin{pmatrix} 1 \\ 0 \end{pmatrix} = \frac{1}{\sqrt{2}} \begin{pmatrix} 1 \\ 1 \end{pmatrix} = |+\rangle$$

$$|1\rangle = \begin{pmatrix} 0 \\ 1 \end{pmatrix} \xrightarrow{\;H\;} \frac{1}{\sqrt{2}} \begin{pmatrix} 1 & 1 \\ 1 & -1 \end{pmatrix} \begin{pmatrix} 0 \\ 1 \end{pmatrix} = \frac{1}{\sqrt{2}} \begin{pmatrix} 1 \\ -1 \end{pmatrix} = |-\rangle$$

where

$$|+\rangle = \frac{1}{\sqrt{2}} \left[\begin{pmatrix} 1 \\ 0 \end{pmatrix} + \begin{pmatrix} 0 \\ 1 \end{pmatrix} \right] = \frac{1}{\sqrt{2}} \begin{pmatrix} 1 \\ 1 \end{pmatrix} \qquad |-\rangle = \frac{1}{\sqrt{2}} \left[\begin{pmatrix} 1 \\ 0 \end{pmatrix} + \begin{pmatrix} 0 \\ -1 \end{pmatrix} \right] = \frac{1}{\sqrt{2}} \begin{pmatrix} 1 \\ -1 \end{pmatrix}$$

In addition, consider two other states that are represented by the notations $|i\rangle$ and $|-i\rangle$ as follows:

$$|i\rangle = \frac{1}{\sqrt{2}}(|0\rangle + i|1\rangle) \qquad |-i\rangle = \frac{1}{\sqrt{2}}(|0\rangle - i|1\rangle)$$

In terms of vector:

$$|i\rangle = \frac{1}{\sqrt{2}} \left[\begin{pmatrix} 1 \\ 0 \end{pmatrix} + \begin{pmatrix} 0 \\ i \end{pmatrix} \right] = \frac{1}{\sqrt{2}} \begin{pmatrix} 1 \\ i \end{pmatrix} \qquad |-i\rangle = \frac{1}{\sqrt{2}} \left[\begin{pmatrix} 1 \\ 0 \end{pmatrix} + \begin{pmatrix} 0 \\ -i \end{pmatrix} \right] = \frac{1}{\sqrt{2}} \begin{pmatrix} 1 \\ -i \end{pmatrix}$$

All these six states, $|0\rangle$, $|1\rangle$, $|+\rangle$, $|-\rangle$, $|i\rangle$, and $|-i\rangle$, can be obtained from each other using quantum gates such as X gate and H gate as well as Y and Z gates, which are presented in Section 2.3.4. These gates are obtained from each other using the Bloch sphere, presented in Section 2.4, and using different transformations and rotations over the x, y, and z axes.

2.3.3 GATE I (IDENTITY)

The identity gate, usually written as I, is basis independent and does not modify the quantum state.

$$I = \begin{pmatrix} 1 & 0 \\ 0 & 1 \end{pmatrix}$$

$$\left\{ \begin{array}{ccc} |0\rangle & \boxed{I} & |0\rangle \\ |1\rangle & \boxed{I} & |1\rangle \\ |\psi\rangle & \boxed{I} & |\psi\rangle \end{array} \right.$$

In general, Gate I is represented only by a wire between two gates or two states:

$$\left\{ \begin{array}{ccc} |0\rangle & \underline{\hspace{2cm}} & |0\rangle \\ |1\rangle & \underline{\hspace{2cm}} & |1\rangle \\ |\psi\rangle & \underline{\hspace{2cm}} & |\psi\rangle \end{array} \right.$$

2.3.4 PAULI-X, Y, AND Z GATES

X, Y, and Z gates, called **Pauli's gates**, act on a single qubit. In the following, these gates is presented by their action on the basic quantic states $|0\rangle$ and $|1\rangle$, as well as by their matrices. The X gate is presented in detail in Section 2.3.1.

2.3.4.1 Gate X (NOT)

The Pauli X gate is the quantum equivalent of the NOT gate for classical computers with respect to the standard basis $|0\rangle$ and $|1\rangle$.

$$X = \begin{pmatrix} 0 & 1 \\ 1 & 0 \end{pmatrix}$$

$$\left\{ \begin{array}{ccc} |0\rangle & \boxed{X} & |1\rangle \\ |1\rangle & \boxed{X} & |0\rangle \end{array} \right.$$

The X gate applied to the other states leads to the following states:

$$\left\{ \begin{array}{ccc} |+\rangle & \boxed{X} & |+\rangle \\ |-\rangle & \boxed{X} & -|-\rangle \\ |i\rangle & \boxed{X} & i|-i\rangle \\ |-i\rangle & \boxed{X} & -i|i\rangle \end{array} \right.$$

2.3.4.2   Gate Y

The Pauli Y gate maps $|0\rangle$ to $i|1\rangle$ and $|1\rangle$ to $-i|0\rangle$.

$$-\boxed{Y}- \qquad Y = \begin{pmatrix} 0 & -i \\ i & 0 \end{pmatrix}$$

$$\left\{ \begin{array}{ccc} |0\rangle & -\boxed{Y}- & i|1\rangle \\ |1\rangle & -\boxed{Y}- & -i|0\rangle \end{array} \right.$$

The Y gate applied to the other states leads to the following states:

$$\left\{ \begin{array}{ccc} |+\rangle & -\boxed{Y}- & -i|-\rangle \\ |-\rangle & -\boxed{Y}- & i|+\rangle \\ |i\rangle & -\boxed{Y}- & |i\rangle \\ |-i\rangle & -\boxed{Y}- & -|-i\rangle \end{array} \right.$$

2.3.4.3   Gate Z

The Pauli Z leaves the basis state $|0\rangle$ unchanged and maps $|1\rangle$ to $-|1\rangle$. Due to this nature, Pauli Z is sometimes called phase-flip.

$$-\boxed{Z}- \qquad Z = \begin{pmatrix} 1 & 0 \\ 0 & -1 \end{pmatrix}$$

$$\left\{ \begin{array}{ccc} |0\rangle & -\boxed{Z}- & |0\rangle \\ |1\rangle & -\boxed{Z}- & -|1\rangle \end{array} \right.$$

The Z gate applied to the other states leads to the following states:

$$\left\{ \begin{array}{ccc} |+\rangle & -\boxed{Z}- & |-\rangle \\ |-\rangle & -\boxed{Z}- & |+\rangle \\ |i\rangle & -\boxed{Z}- & |-i\rangle \\ |-i\rangle & -\boxed{Z}- & |i\rangle \end{array} \right.$$

2.3.5   GATE $\sqrt{NOT}$

It is also possible to consider the gate $M = \sqrt{NOT}$ defined as follows:

$$-\boxed{\sqrt{NOT}}- \qquad M = \begin{pmatrix} 1+i & 1-i \\ 1-i & 1+i \end{pmatrix}$$

$$\begin{cases} |0\rangle \quad -\boxed{\sqrt{NOT}} \longrightarrow \quad \frac{1+i}{2}|0\rangle - \frac{1-i}{2}|1\rangle \\[2em] |1\rangle \quad -\boxed{\sqrt{NOT}} \longrightarrow \quad \frac{1-i}{2}|0\rangle - \frac{1+i}{2}|1\rangle \end{cases}$$

2.3.6 1-QUBIT GATES: SUMMARY

The different transformations of a qubit by applying the classical quantum gates are given in Table 2.1.

Table 2.1
1-Qubit Gates

State	Gate X	Gate Y	Gate Z
$|0\rangle$	$|1\rangle$	$i|1\rangle$	$|0\rangle$
$|1\rangle$	$|0\rangle$	$-i|0\rangle$	$-|1\rangle$
$|+\rangle$	$|+\rangle$	$-i|-\rangle$	$|-\rangle$
$|-\rangle$	$-|-\rangle$	$i|+\rangle$	$|+\rangle$
$|i\rangle$	$i|-i\rangle$	$|i\rangle$	$|-i\rangle$
$|-i\rangle$	$-i|i\rangle$	$-|-i\rangle$	$|i\rangle$

2.4 BLOCH SPHERE

A Bloch sphere (see Figure 2.2(a)) can be used to represent qubits in a simple way. It can help to easily identify the transformation of the qubit states. This sphere is represented in a 3D space with three axes x, y, and z. Applying any Pauli's gate (i.e., X, Y, or Z) to any qubit performs a rotation of π over the corresponding axis (x for X gate, y for Y gate and z for Z gate). As an example, the qubit $|0\rangle$ is represented by the red vector with the norm 1 and lying on the upper side of the z-axis. The application of the X gate to this qubit leads to a rotation of π over the x-axis. The result is the qubit $|1\rangle$. The application of the Y gate to $|0\rangle$ leads to the same result, i.e., $|1\rangle$. In the same way, applying either the Z gate or the Y gate to $|-\rangle$ results in the $|+\rangle$ state. This $|+\rangle$ state can also be obtained from the $|0\rangle$ state by applying the H gate to the qubit $|0\rangle$, which means that the Hadamard H gate is transforming any qubit by performing a rotation of π over the v-axis as shown by Figure 2.2(b).

The application of the gates is reversible. This means that applying twice the same gate to a qubit leads to the same qubit. In other terms, applying twice any gate to $|\psi\rangle$ leads to a qubit $|\psi\rangle$. Therefore, applying twice the Gate X to $|0\rangle$ leads to the qubit $|0\rangle$ as shown in

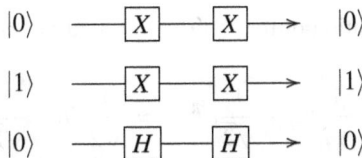

$$|0\rangle \quad -\boxed{X}-\boxed{X} \longrightarrow \quad |0\rangle$$

$$|1\rangle \quad -\boxed{X}-\boxed{X} \longrightarrow \quad |1\rangle$$

$$|0\rangle \quad -\boxed{H}-\boxed{H} \longrightarrow \quad |0\rangle$$

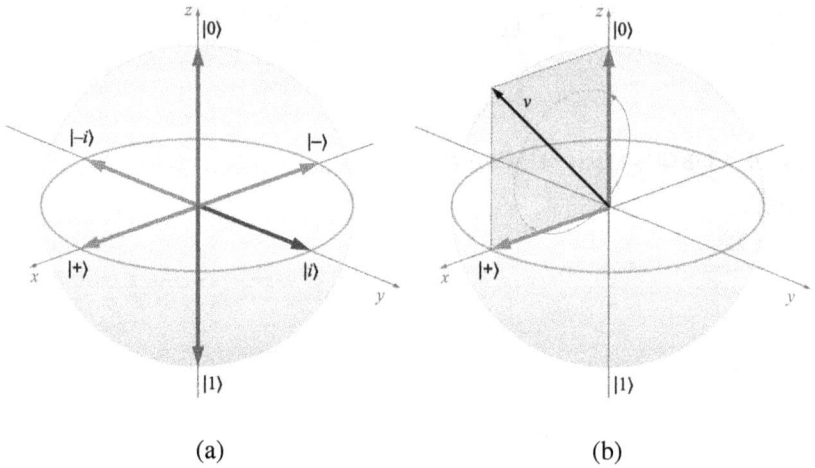

(a)   (b)

Figure 2.2   (a) Bloch sphere, (b) v-axis.

2.5   PHASE GATES

The phase gate P_φ of a state ψ represents its rotation around the z-axis (on the Bloch sphere) by φ radians (cf. Figure 2.3).

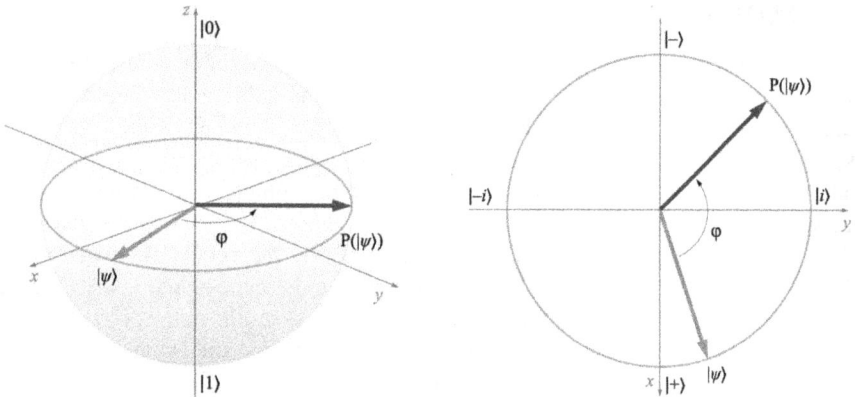

Figure 2.3   Bloch sphere 3 (phase gate transformation).

We can then write:

$$P_\varphi(|\psi\rangle) = \alpha|0\rangle + e^{i\varphi}\beta|1\rangle$$

For example, the rotation of the state $|+\rangle$ around the z-axis by π radians results in:

$$P_\pi(|+\rangle) = \frac{1}{\sqrt{2}}|0\rangle + \frac{1}{\sqrt{2}}e^{i\pi}|1\rangle = \frac{1}{\sqrt{2}}|0\rangle - \frac{1}{\sqrt{2}}|1\rangle = |-\rangle$$

Note that:

$$e^{i\pi} = i^2 = -1$$

2.5.1 PHASE GATE (R_K)

The R_k gate represents rotation around the z-axis of $\frac{2\pi}{2^k}$ radian.

$$R_k = \begin{pmatrix} 1 & 0 \\ 0 & e^{i\frac{2\pi}{2^k}} \end{pmatrix} = \begin{pmatrix} 1 & 0 \\ 0 & (e^{i\pi})^{\frac{2}{2^k}} \end{pmatrix} = \begin{pmatrix} 1 & 0 \\ 0 & -1^{2/2^k} \end{pmatrix}$$

2.5.2 PHASE GATE (S)

The S gate, also known as the phase gate, is referred to as the Z90 because it represents a 90-degree rotation ($\frac{\pi}{2}$ radians) around the z-axis. Instead of going halfway around the Bloch sphere, we only go a quarter (a 90-degree turn). The T gate is related to the S gate by the relationship $S = T^2$. It is equivalent to gate R_2.

$$S = \begin{pmatrix} 1 & 0 \\ 0 & e^{i\frac{\pi}{2}} \end{pmatrix} = \begin{pmatrix} 1 & 0 \\ 0 & i \end{pmatrix}$$

$$\begin{cases} |+\rangle & \boxed{S} & |i\rangle \\ |i\rangle & \boxed{S} & |-\rangle \\ |-\rangle & \boxed{S} & |-i\rangle \\ |-i\rangle & \boxed{S} & |+\rangle \end{cases}$$

These states are presented in the Bloch sphere of Figure 2.4(a)

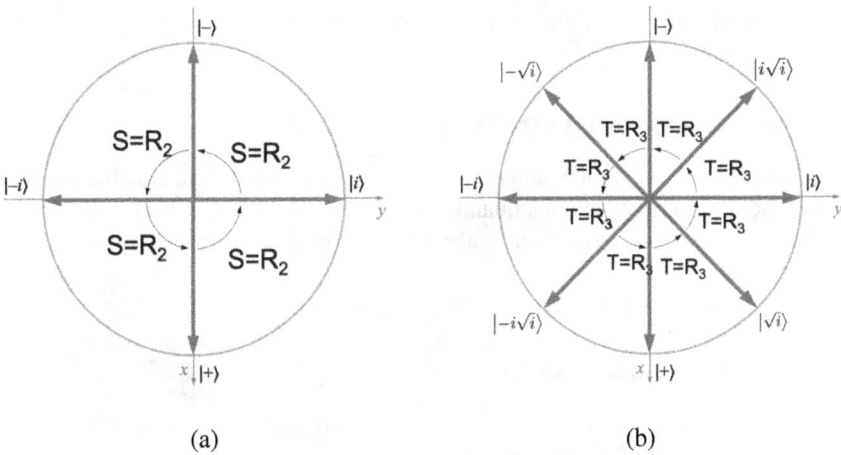

(a)   (b)

Figure 2.4 Bloch sphere: (a) S gate rotations, (b) T gate rotations.

Table 2.2

1-Qubit S Gate

State	Gate S
$|+\rangle$	$|i\rangle$
$|i\rangle$	$|-\rangle$
$|-\rangle$	$|-i\rangle$
$|-i\rangle$	$|+\rangle$

2.5.3 PHASE GATE T

The T gate performs a quarter turn ($\frac{\pi}{4}$ radians) of the Z gate and half of the S gate. It is equivalent to gate R_3.

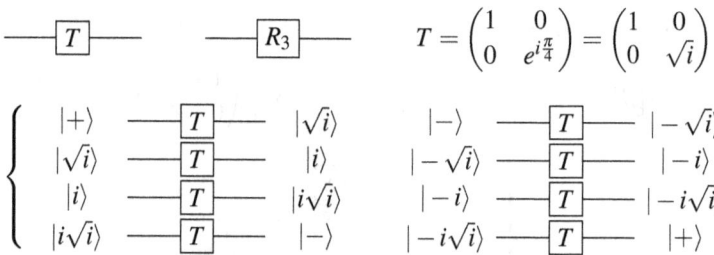

$$T = \begin{pmatrix} 1 & 0 \\ 0 & e^{i\frac{\pi}{4}} \end{pmatrix} = \begin{pmatrix} 1 & 0 \\ 0 & \sqrt{i} \end{pmatrix}$$

$$\begin{cases} |+\rangle \xrightarrow{T} |\sqrt{i}\rangle & |-\rangle \xrightarrow{T} |-\sqrt{i}\rangle \\ |\sqrt{i}\rangle \xrightarrow{T} |i\rangle & |-\sqrt{i}\rangle \xrightarrow{T} |-i\rangle \\ |i\rangle \xrightarrow{T} |i\sqrt{i}\rangle & |-i\rangle \xrightarrow{T} |-i\sqrt{i}\rangle \\ |i\sqrt{i}\rangle \xrightarrow{T} |-\rangle & |-i\sqrt{i}\rangle \xrightarrow{T} |+\rangle \end{cases}$$

These states are presented in the Bloch sphere of Figure 2.4(b).

2.5.4 PHASE GATES: SUMMARY

The different transformations of a qubit by applying the phase gates are given in Tables 2.2 and 2.3.

2.6 MEASUREMENT AND CIRCUIT

A **Measurement** is a fundamental element of a quantum circuit. It is the only moment when information about a quantum state $|\psi\rangle$ is obtained. It also marks the end of the qubit, as the measurement only returns 0 or 1 and irreversibly disturbs the quantum state.

A **Circuit** is composed of a sequence of gates, read from left to right.

2.6.1 X GATE MEASUREMENT

The following illustration shows an example of a circuit composed of an X gate, i.e., a NOT gate, followed by a measurement gate.

Table 2.3
1-Qubit T Gate

State	Gate T
$|+\rangle$	$|\sqrt{i}\rangle$
$|\sqrt{i}\rangle$	$|i\rangle$
$|i\rangle$	$|i\sqrt{i}\rangle$
$|i\sqrt{i}\rangle$	$|-\rangle$
$|-\rangle$	$|-\sqrt{i}\rangle$
$|-\sqrt{i}\rangle$	$|-i\rangle$
$|-i\rangle$	$|-i\sqrt{i}\rangle$
$|-i\sqrt{i}\rangle$	$|+\rangle$

If the input is $|0\rangle$, then $X(|0\rangle) = |1\rangle$, so the measured output is 1 (with a probability of 1)

$$|0\rangle \quad \boxed{X} \boxed{\nearrow} \quad 1$$

If the input is $|1\rangle$, then $X(|1\rangle) = |0\rangle$, so the measured output is 0 (with a probability of 1)

$$|1\rangle \quad \boxed{X} \boxed{\nearrow} \quad 0$$

If the input is the state $|\psi\rangle = \alpha|0\rangle + \beta|1\rangle$ (with $|\alpha|^2 + |\beta|^2 = 1$), then $X(|\psi\rangle) = \beta|0\rangle + \alpha|1\rangle$. The measurement thus yields 0 with a probability of $|\beta|^2$ and 1 with a probability of $|\alpha|^2$.

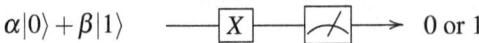

$$\alpha|0\rangle + \beta|1\rangle \quad \boxed{X} \boxed{\nearrow} \quad 0 \text{ or } 1$$

2.6.2 H GATE MEASUREMENT

The following illustrations shows another example of a circuit composed of an H gate followed by a measurement gate.

$$\boxed{H} \boxed{\nearrow}$$

If the input is $|0\rangle$, then $H(|0\rangle) = \frac{1}{\sqrt{2}}|0\rangle + \frac{1}{\sqrt{2}}|1\rangle$, so the measured output is 1 (with a probability of $\frac{1}{2}$) and 0 (with a probability of $\frac{1}{2}$)

$$|0\rangle \quad \boxed{H} \boxed{\nearrow} \quad 0 \text{ or } 1 \text{ (with a probability of } \frac{1}{2} \text{ each)}$$

If the input is $|1\rangle$, then $H(|1\rangle) = \frac{1}{\sqrt{2}}|0\rangle - \frac{1}{\sqrt{2}}|1\rangle$, so the measured output is 1 (with a probability of $\frac{1}{2}$) and 0 (with a probability of $\frac{1}{2}$):

$$|1\rangle \quad \boxed{H} \quad \boxed{\angle} \longrightarrow \qquad 0 \text{ or } 1 \text{ (with a probability of } \tfrac{1}{2} \text{ each)}$$

However, if the input is the state $|+\rangle = \frac{1}{\sqrt{2}}|0\rangle + \frac{1}{\sqrt{2}}|1\rangle$, then:

$$
\begin{aligned}
H(|+\rangle) &= H\left(\frac{1}{\sqrt{2}}|0\rangle + \frac{1}{\sqrt{2}}|1\rangle\right) \\
&= \frac{1}{\sqrt{2}}H(|0\rangle) + \frac{1}{\sqrt{2}}H(|1\rangle) \\
&= \frac{1}{\sqrt{2}}\left(\frac{1}{\sqrt{2}}|0\rangle + \frac{1}{\sqrt{2}}|1\rangle\right) + \frac{1}{\sqrt{2}}\left(\frac{1}{\sqrt{2}}|0\rangle - \frac{1}{\sqrt{2}}|1\rangle\right) \\
&= \frac{1}{2}|0\rangle + \frac{1}{2}|1\rangle + \frac{1}{2}|0\rangle - \frac{1}{2}|1\rangle \\
&= \frac{1}{2}|0\rangle + \frac{1}{2}|0\rangle \\
&= |0\rangle
\end{aligned}
\tag{2.3}
$$

which means that:
$$|+\rangle \xmapsto{\ H\ } |0\rangle$$

Thus, for this input, the measured output returns 0 with a probability of 1.

$$|+\rangle \quad \boxed{H} \quad \boxed{\angle} \longrightarrow \qquad 0$$

Finally, we can conclude that the measured output of $|-\rangle \xmapsto{\ H\ } |1\rangle$ returns 1 with a probability of 1.

$$|-\rangle \quad \boxed{H} \quad \boxed{\angle} \longrightarrow \qquad 1$$

2.7   A 2-QUBIT

2.7.1   SUPERPOSITION

Two combined qubits are in a quantum state $|\psi\rangle$, referred to as a 2-qubit state, are defined by the following superposition:

$$|\psi\rangle = \alpha|00\rangle + \beta|01\rangle + \gamma|10\rangle + \delta|11\rangle$$

where $\alpha, \beta, \gamma, \delta \in \mathbb{C}$, with the normalisation convention:

$$|\alpha|^2 + |\beta|^2 + |\gamma|^2 + |\delta|^2 = 1$$

The measurement of a 2-qubit lead to two classical bits:

$$
\begin{aligned}
&00 \text{ with a probability } |\alpha|^2\\
&01 \text{ with a probability } |\beta|^2\\
&10 \text{ with a probability } |\gamma|^2\\
&11 \text{ with a probability } |\delta|^2
\end{aligned}
\tag{2.4}
$$

Note the remarkable difference from classical computing: with two classical bits, only one of the four states 00, 01, 10, or 11 can be encoded. However, with a 2-qubit state, all four states can be encoded simultaneously.

$|00\rangle, |01\rangle, |10\rangle$, and $|11\rangle$ are vectors of a basis in a four-dimensional space, given as follows:

$$
|00\rangle = \begin{pmatrix} 1 \\ 0 \\ 0 \\ 0 \end{pmatrix} \quad
|01\rangle = \begin{pmatrix} 0 \\ 1 \\ 0 \\ 0 \end{pmatrix} \quad
|10\rangle = \begin{pmatrix} 0 \\ 0 \\ 1 \\ 0 \end{pmatrix} \quad
|11\rangle = \begin{pmatrix} 0 \\ 0 \\ 0 \\ 1 \end{pmatrix}
\tag{2.5}
$$

Therefore, $|\psi\rangle$ can be written as a vector in $\mathbb{C}^4$ as follows:

$$
|\psi\rangle = \alpha \begin{pmatrix} 1 \\ 0 \\ 0 \\ 0 \end{pmatrix}
+ \beta \begin{pmatrix} 0 \\ 1 \\ 0 \\ 0 \end{pmatrix}
+ \gamma \begin{pmatrix} 0 \\ 0 \\ 1 \\ 0 \end{pmatrix}
+ \delta \begin{pmatrix} 0 \\ 0 \\ 0 \\ 1 \end{pmatrix}
= \begin{pmatrix} \alpha \\ \beta \\ \gamma \\ \delta \end{pmatrix}
$$

As an example:

$$
|\psi\rangle = \frac{1}{\sqrt{6}}|00\rangle + \frac{i}{\sqrt{6}}|10\rangle + \frac{1+i}{\sqrt{3}}|11\rangle
\tag{2.6}
$$

is a 2-qubit with the norm 1. Its measurement gives:

$$
\begin{aligned}
&00 \text{ with a probability } 1/6\\
&01 \text{ with a probability } 0\\
&10 \text{ with a probability } 1/6\\
&11 \text{ with a probability } 2/3
\end{aligned}
\tag{2.7}
$$

One can also represent the basic states using tensor product ($\otimes$) formulas:

$$
|00\rangle = |0\rangle \otimes |0\rangle \quad
|01\rangle = |0\rangle \otimes |1\rangle \quad
|10\rangle = |1\rangle \otimes |0\rangle \quad
|11\rangle = |1\rangle \otimes |1\rangle
$$

The tensor product is calculated as follows, using vector representation:

$$
\begin{pmatrix} x \\ y \end{pmatrix} \otimes \begin{pmatrix} a \\ b \\ c \end{pmatrix}
= \begin{pmatrix} x\begin{pmatrix} a \\ b \\ c \end{pmatrix} \\ y\begin{pmatrix} a \\ b \\ c \end{pmatrix} \end{pmatrix}
= \begin{pmatrix} xa \\ xb \\ xc \\ ya \\ yb \\ yc \end{pmatrix}
$$

For the case of $|01\rangle$, it can be written as:

$$|01\rangle = |0\rangle \otimes |1\rangle = \begin{pmatrix} 1 \\ 0 \end{pmatrix} \otimes \begin{pmatrix} 0 \\ 1 \end{pmatrix} = \begin{pmatrix} 1\begin{pmatrix} 0 \\ 1 \end{pmatrix} \\ 0\begin{pmatrix} 0 \\ 1 \end{pmatrix} \end{pmatrix} = \begin{pmatrix} 1 \times 0 \\ 1 \times 1 \\ 0 \times 0 \\ 0 \times 1 \end{pmatrix} = \begin{pmatrix} 0 \\ 1 \\ 0 \\ 0 \end{pmatrix}$$

which is already presented by Equation (2.5).

Equation (2.6) can then be rewritten as:

$$|\psi\rangle = \begin{pmatrix} 1/6 \\ 0 \\ 1/6 \\ 2/3 \end{pmatrix}$$

2.7.2 GATE CNOT (CX)

The CNOT gate, also called CX gate, is a gate that takes the two qubits $|x\rangle$ and $|y\rangle$ as input and returns the two qubits $|x\rangle$ and $|x \oplus y\rangle$ as output.

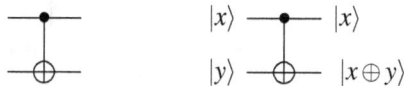

$$|x\rangle \quad\text{—•—}\quad |x\rangle$$
$$|y\rangle \quad\text{—⊕—}\quad |x \oplus y\rangle$$

In this gate, the first qubit remains the same at the output. However, the second qubit, which depends on the first one, is equal to $|x \oplus y\rangle$. This can be explained using the following rule:

- if the first qubit is $|0\rangle$, the second qubit remains the same at the output:

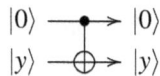

$$|0\rangle \quad\text{—•→}\quad |0\rangle$$
$$|y\rangle \quad\text{—⊕→}\quad |y\rangle$$

- if the first qubit is $|1\rangle$, the second qubit is reversed based on the Gate X:

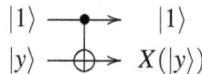

$$|1\rangle \quad\text{—•→}\quad |1\rangle$$
$$|y\rangle \quad\text{—⊕→}\quad X(|y\rangle)$$

The following cases show the four basic quantum states:

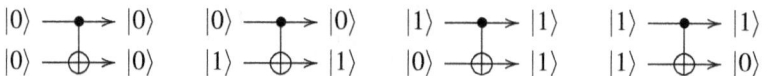

$$|0\rangle \to |0\rangle \qquad |0\rangle \to |0\rangle \qquad |1\rangle \to |1\rangle \qquad |1\rangle \to |1\rangle$$
$$|0\rangle \to |0\rangle \qquad |1\rangle \to |1\rangle \qquad |0\rangle \to |1\rangle \qquad |1\rangle \to |0\rangle$$

The CNOT gate can be considered as an instruction "if, then, else": if the first qubit is $|0\rangle$, then do this; else, do that.

The states can be written with the notation of 2-qubits as follows:

$$|00\rangle \xrightarrow{CNOT} |00\rangle \quad |01\rangle \xrightarrow{CNOT} |01\rangle \quad |10\rangle \xrightarrow{CNOT} |11\rangle \quad |11\rangle \xrightarrow{CNOT} |10\rangle$$

The same states can be written using vectors as follows:

$$\begin{pmatrix} 1 \\ 0 \\ 0 \\ 0 \end{pmatrix} \mapsto \begin{pmatrix} 1 \\ 0 \\ 0 \\ 0 \end{pmatrix} \quad \begin{pmatrix} 0 \\ 1 \\ 0 \\ 0 \end{pmatrix} \mapsto \begin{pmatrix} 0 \\ 1 \\ 0 \\ 0 \end{pmatrix} \quad \begin{pmatrix} 0 \\ 0 \\ 1 \\ 0 \end{pmatrix} \mapsto \begin{pmatrix} 0 \\ 0 \\ 0 \\ 1 \end{pmatrix} \quad \begin{pmatrix} 0 \\ 0 \\ 0 \\ 1 \end{pmatrix} \mapsto \begin{pmatrix} 0 \\ 0 \\ 1 \\ 0 \end{pmatrix}$$

The matrix M of the transformation CNOT is given as follows:

$$M = \begin{pmatrix} 1 & 0 & 0 & 0 \\ 0 & 1 & 0 & 0 \\ 0 & 0 & 0 & 1 \\ 0 & 0 & 1 & 0 \end{pmatrix}$$

The CNOT gate transforms a vector representing a 2-qubit system through multiplication by this matrix M:

$$\begin{pmatrix} \alpha \\ \beta \\ \gamma \\ \delta \end{pmatrix} \xrightarrow{CNOT} \begin{pmatrix} 1 & 0 & 0 & 0 \\ 0 & 1 & 0 & 0 \\ 0 & 0 & 0 & 1 \\ 0 & 0 & 1 & 0 \end{pmatrix} \begin{pmatrix} \alpha \\ \beta \\ \gamma \\ \delta \end{pmatrix} = \begin{pmatrix} \alpha \\ \beta \\ \delta \\ \gamma \end{pmatrix}$$

2.7.3 SWAP GATE

The SWAP gate exchanges two qubits. It is represented as:

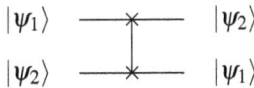

The SWAP gate is equivalent to the following circuit:

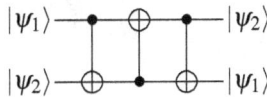

The matrix M of the transformation SWAP is given as follows:

$$M = \begin{pmatrix} 1 & 0 & 0 & 0 \\ 0 & 0 & 1 & 0 \\ 0 & 1 & 0 & 0 \\ 0 & 0 & 0 & 1 \end{pmatrix}$$

2.7.4 BELL STATE (QUANTUM ENTANGLEMENT)

To obtain the **Bell State**, consider the following circuit, which is composed of the Hadamard gate followed by the CNOT gate:

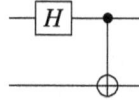

Then, from the output $|00\rangle$, the Bell State is obtained at the output. This can be explained by the fact that the first input $|0\rangle$ is transformed by the Hadamard gate $H(|0\rangle) = \frac{1}{\sqrt{2}}|0\rangle + \frac{1}{\sqrt{2}}|1\rangle$. This means that the first output can be $|0\rangle$ with the probability 0.5 and can be $|1\rangle$ with the same probability 0.5.

Thus, if the first output is equal to $|0\rangle$, then the second output remains the same. Since this one is equal to $|0\rangle$, then its output is $|0\rangle$. Therefore, if the first output is $|0\rangle$, then the second output is also $|0\rangle$.

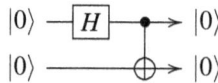

On the other hand, if the first output is equal to $|1\rangle$, then the X gate is applied to the second input. Since this input is $|0\rangle$ the X gate transforms it to $|1\rangle$. Therefore, if the first output is $|1\rangle$, then the second output is $|1\rangle$ too.

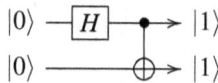

The **Bell State** $|\Phi^+\rangle$ can be written as:

$$|\Phi^+\rangle = \frac{1}{\sqrt{2}}|00\rangle + \frac{1}{\sqrt{2}}|11\rangle$$

In terms of vectors, we can write:

$$|\Phi^+\rangle = \frac{1}{\sqrt{2}}\begin{pmatrix} 1 \\ 0 \\ 0 \\ 0 \end{pmatrix} + \frac{1}{\sqrt{2}}\begin{pmatrix} 0 \\ 0 \\ 0 \\ 1 \end{pmatrix} = \frac{1}{\sqrt{2}}\begin{pmatrix} 1 \\ 0 \\ 0 \\ 1 \end{pmatrix}$$

Measuring this state leads to:

$$\begin{aligned} &00 \text{ with a probability } 1/2 \\ &11 \text{ with a probability } 1/2 \end{aligned} \tag{2.8}$$

The other outputs 01 and 10 have a probability 0 to occur.

Consider the following analogy:

- A qubit is somewhat like a coin tossed in the air. As long as the coin is spinning in the air, *heads* and *tails* have an equal chance of occurring. It is only when the coin has landed that one can read the result (this is the *measurement* part), and then the result is definitively fixed to either *heads* or *tails*.
- A 2-qubit system, the combination of two qubits, is like tossing two coins in the air simultaneously. The four possible outcomes, *heads/heads*, *heads/tails*, *tails/heads*, or *tails/tails*, are all possible.
- The Bell State is like tossing two coins linked together in the air. The result can only be either *heads/heads* or *tails/tails*. This phenomenon is called **Quantum Entanglement**.

2.7.5 H GATE (WITH 2-QUBITS)

The Hadamard gate with one input is defined in Section 2.3.2. In this section, we present the same gate with two inputs as:

$$|\psi\rangle = |\psi_1 \psi_2\rangle \quad \xrightarrow{\;\;2\;\;} \boxed{H^{\otimes 2}} \longrightarrow H^{\otimes 2}(|\psi\rangle)$$

Or, as:

$$|\psi_1\rangle \longrightarrow \boxed{H} \longrightarrow$$
$$H^{\otimes 2}(|\psi\rangle)$$
$$|\psi_2\rangle \longrightarrow \boxed{H} \longrightarrow$$

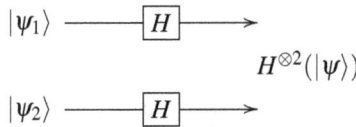

It is defined as:

a. For $|00\rangle$,

$$|00\rangle = |0\rangle \otimes |0\rangle \xrightarrow{\;H\;} H^{\otimes 2}(|00\rangle) = H(|0\rangle) \otimes H(|0\rangle) =$$

$$\left(\frac{1}{\sqrt{2}}|0\rangle + \frac{1}{\sqrt{2}}|1\rangle\right) \otimes \left(\frac{1}{\sqrt{2}}|0\rangle + \frac{1}{\sqrt{2}}|1\rangle\right) =$$

$$\frac{1}{\sqrt{2}}(|0\rangle + |1\rangle) \otimes \frac{1}{\sqrt{2}}(|0\rangle + |1\rangle) =$$

$$\frac{1}{2}(|0\rangle + |1\rangle) \otimes (|0\rangle + |1\rangle) =$$

$$\frac{1}{2}(|0\rangle \otimes |0\rangle + |0\rangle \otimes |1\rangle + |1\rangle \otimes |0\rangle + |1\rangle \otimes |1\rangle) =$$

$$\frac{1}{2}(|00\rangle + |01\rangle + |10\rangle + |11\rangle)$$

Thus,

$$H^{\otimes 2}(|00\rangle) = \frac{1}{2}(|00\rangle + |01\rangle + |10\rangle + |11\rangle)$$

b. And for $|01\rangle$,

$$|01\rangle = |0\rangle \otimes |1\rangle \xmapsto{\;H\;} H^{\otimes 2}(|01\rangle) = H(|0\rangle) \otimes H(|1\rangle) =$$

$$\left(\frac{1}{\sqrt{2}}|0\rangle + \frac{1}{\sqrt{2}}|1\rangle \right) \otimes \left(\frac{1}{\sqrt{2}}|0\rangle - \frac{1}{\sqrt{2}}|1\rangle \right) =$$

$$\frac{1}{\sqrt{2}}(|0\rangle + |1\rangle) \otimes \frac{1}{\sqrt{2}}(|0\rangle - |1\rangle) =$$

$$\frac{1}{2}(|0\rangle + |1\rangle) \otimes (|0\rangle - |1\rangle) =$$

$$\frac{1}{2}(|0\rangle \otimes |0\rangle - |0\rangle \otimes |1\rangle + |1\rangle \otimes |0\rangle - |1\rangle \otimes |1\rangle) =$$

$$\frac{1}{2}(|00\rangle - |01\rangle + |10\rangle - |11\rangle)$$

Thus,

$$H^{\otimes 2}(|01\rangle) = \frac{1}{2}(|00\rangle - |01\rangle + |10\rangle - |11\rangle)$$

c. And for $|10\rangle$,

$$|10\rangle = |1\rangle \otimes |0\rangle \xmapsto{\;H\;} H^{\otimes 2}(|10\rangle) = H(|1\rangle) \otimes H(|0\rangle) =$$

$$\left(\frac{1}{\sqrt{2}}|0\rangle - \frac{1}{\sqrt{2}}|1\rangle \right) \otimes \left(\frac{1}{\sqrt{2}}|0\rangle + \frac{1}{\sqrt{2}}|1\rangle \right) =$$

$$\frac{1}{\sqrt{2}}(|0\rangle - |1\rangle) \otimes \frac{1}{\sqrt{2}}(|0\rangle + |1\rangle) =$$

$$\frac{1}{2}(|0\rangle - |1\rangle) \otimes (|0\rangle + |1\rangle) =$$

$$\frac{1}{2}(|0\rangle \otimes |0\rangle + |0\rangle \otimes |1\rangle - |1\rangle \otimes |0\rangle - |1\rangle \otimes |1\rangle) =$$

$$\frac{1}{2}(|00\rangle + |01\rangle - |10\rangle - |11\rangle)$$

Thus,

$$H^{\otimes 2}(|10\rangle) = \frac{1}{2}(|00\rangle + |01\rangle - |10\rangle - |11\rangle)$$

d. And for $|11\rangle$,

$$|11\rangle = |1\rangle \otimes |1\rangle \xmapsto{\;H\;} H^{\otimes 2}(|11\rangle) = H(|1\rangle) \otimes H(|1\rangle) =$$

$$\left(\frac{1}{\sqrt{2}}|0\rangle - \frac{1}{\sqrt{2}}|1\rangle \right) \otimes \left(\frac{1}{\sqrt{2}}|0\rangle - \frac{1}{\sqrt{2}}|1\rangle \right) =$$

$$\frac{1}{\sqrt{2}}(|0\rangle - |1\rangle) \otimes \frac{1}{\sqrt{2}}(|0\rangle - |1\rangle) =$$

$$\frac{1}{2}(|0\rangle - |1\rangle) \otimes (|0\rangle - |1\rangle) =$$

$$\frac{1}{2}(|0\rangle \otimes |0\rangle - |0\rangle \otimes |1\rangle - |1\rangle \otimes |0\rangle + |1\rangle \otimes |1\rangle) =$$

$$\frac{1}{2}(|00\rangle - |01\rangle - |10\rangle + |11\rangle)$$

Thus,

$$H^{\otimes 2}(|11\rangle) = \frac{1}{2}(|00\rangle - |01\rangle - |10\rangle + |11\rangle)$$

In terms of vectors, we can write:

$$H^{\otimes 2}(|00\rangle) = H(|0\rangle) \otimes H(|0\rangle)$$

$$\Rightarrow H \begin{pmatrix} 1 \\ 0 \\ 0 \\ 0 \end{pmatrix} = \frac{1}{2}\left[\begin{pmatrix} 1 \\ 1 \end{pmatrix} \otimes \begin{pmatrix} 1 \\ 1 \end{pmatrix}\right] = \frac{1}{2}\begin{pmatrix} 1 \\ 1 \\ 1 \\ 1 \end{pmatrix}$$

$$H^{\otimes 2}(|01\rangle) = H(|0\rangle) \otimes H(|1\rangle)$$

$$\Rightarrow H \begin{pmatrix} 0 \\ 1 \\ 0 \\ 0 \end{pmatrix} = \frac{1}{2}\left[\begin{pmatrix} 1 \\ 1 \end{pmatrix} \otimes \begin{pmatrix} 1 \\ -1 \end{pmatrix}\right] = \frac{1}{2}\begin{pmatrix} 1 \\ -1 \\ 1 \\ -1 \end{pmatrix}$$

$$H^{\otimes 2}(|10\rangle) = H(|1\rangle) \otimes H(|0\rangle)$$

$$\Rightarrow H \begin{pmatrix} 0 \\ 0 \\ 1 \\ 0 \end{pmatrix} = \frac{1}{2}\left[\begin{pmatrix} 1 \\ -1 \end{pmatrix} \otimes \begin{pmatrix} 1 \\ 1 \end{pmatrix}\right] = \frac{1}{2}\begin{pmatrix} 1 \\ 1 \\ -1 \\ -1 \end{pmatrix}$$

$$H^{\otimes 2}(|11\rangle) = H(|1\rangle) \otimes H(|1\rangle)$$

$$\Rightarrow H \begin{pmatrix} 0 \\ 0 \\ 0 \\ 1 \end{pmatrix} = \frac{1}{2}\left[\begin{pmatrix} 1 \\ -1 \end{pmatrix} \otimes \begin{pmatrix} 1 \\ -1 \end{pmatrix}\right] = \frac{1}{2}\begin{pmatrix} 1 \\ -1 \\ -1 \\ 1 \end{pmatrix}$$

The matrix $H^{\otimes 2}$ of the transformation of Hadamard for a 2-qubits is given as follows:

$$H^{\otimes 2} = H \otimes H = \frac{1}{\sqrt{2}}\begin{pmatrix} 1 & 1 \\ 1 & -1 \end{pmatrix} \otimes \frac{1}{\sqrt{2}}\begin{pmatrix} 1 & 1 \\ 1 & -1 \end{pmatrix} = \frac{1}{2}\left(\begin{array}{cc|cc} 1 & 1 & 1 & 1 \\ 1 & -1 & 1 & -1 \\ \hline 1 & 1 & -1 & -1 \\ 1 & -1 & -1 & 1 \end{array}\right)$$

We recall that the tensor product, using matrix representation is calculated as follows:

$$\begin{pmatrix} a & b \\ c & d \end{pmatrix} \otimes \begin{pmatrix} e & f \\ g & h \end{pmatrix} = \left(\begin{array}{c|c} a\begin{pmatrix} e & f \\ g & h \end{pmatrix} & b\begin{pmatrix} e & f \\ g & h \end{pmatrix} \\ \hline c\begin{pmatrix} e & f \\ g & h \end{pmatrix} & d\begin{pmatrix} e & f \\ g & h \end{pmatrix} \end{array}\right)$$

Using this representation, we calculate $H^{\otimes 2}(|\psi\rangle)$, where $|\psi\rangle = \alpha|00\rangle + \beta|01\rangle + \gamma|10\rangle + \delta|11\rangle$ as follows:

$$H^{\otimes 2}|\psi\rangle = \frac{1}{2}\begin{pmatrix} 1 & 1 & 1 & 1 \\ 1 & -1 & 1 & -1 \\ 1 & 1 & -1 & -1 \\ 1 & -1 & -1 & 1 \end{pmatrix}\begin{pmatrix} \alpha \\ \beta \\ \gamma \\ \delta \end{pmatrix} = \frac{1}{2}\begin{pmatrix} \alpha + \beta + \gamma + \delta \\ \alpha - \beta + \gamma - \delta \\ \alpha + \beta - \gamma - \delta \\ \alpha - \beta - \gamma + \delta \end{pmatrix}$$

This can be written as follows:

$$\begin{aligned} H^{\otimes 2}(|\psi\rangle) &= H(\alpha|00\rangle + \beta|01\rangle + \gamma|10\rangle + \delta|11\rangle) \\ &= \alpha H(|00\rangle) + \beta H(|01\rangle) + \gamma H(|10\rangle) + \delta H(|11\rangle) \\ &= \alpha\frac{1}{2}(|00\rangle + |01\rangle + |10\rangle + |11\rangle) + \\ &\quad \beta\frac{1}{2}(|00\rangle - |01\rangle + |10\rangle - |11\rangle) + \\ &\quad \gamma\frac{1}{2}(|00\rangle + |01\rangle - |10\rangle - |11\rangle) + \\ &\quad \delta\frac{1}{2}(|00\rangle - |01\rangle - |10\rangle + |11\rangle) \\ &= \left(\frac{\alpha + \beta + \gamma + \delta}{2}|00\rangle\right) + \left(\frac{\alpha - \beta + \gamma - \delta}{2}|01\rangle\right) + \\ &\quad \left(\frac{\alpha + \beta - \gamma - \delta}{2}|10\rangle\right) + \left(\frac{\alpha - \beta - \gamma + \delta}{2}|11\rangle\right) \end{aligned}$$

Thus,

$$H^{\otimes 2}(|\psi\rangle) = \left(\frac{\alpha + \beta + \gamma + \delta}{2}|00\rangle\right) + \left(\frac{\alpha - \beta + \gamma - \delta}{2}|01\rangle\right) + \\ \left(\frac{\alpha + \beta - \gamma - \delta}{2}|10\rangle\right) + \left(\frac{\alpha - \beta - \gamma + \delta}{2}|11\rangle\right)$$

2.8   AN N-QUBIT

2.8.1   SUPERPOSITION

An n-qubit is a quantum state $|\psi\rangle$ defined by the following superposition:

$$|\psi\rangle = \alpha_0|00\ldots0\rangle + \alpha_1|00\ldots1\rangle + \ldots + \alpha_{2^n-1}|11\ldots1\rangle$$

where $\alpha_0, \alpha_1, \ldots, \alpha_{2^n-1}, \in \mathbb{C}$, with the normalisation convention:

$$|\alpha_0|^2 + |\alpha_1|^2 + \cdots + |\alpha_{2^n-1}|^2 = 1$$

We can also write $|\psi\rangle$ as follows:

$$|\psi\rangle = \alpha \sum_{x \in \{0,1\}^n} |x\rangle$$

where, $\alpha = (\alpha_1, \alpha_2, ..., \alpha_{2^n-1})$

An n-qubit, therefore, has 2^n coefficients. This is the entire power of quantum computing: the combination of n qubits leads to the superposition of 2^n basic states. Working with an n-qubit corresponds to simultaneously working with all 2^n classical n-bit states $00...0$, $00...1$, ..., $11...1$, while classical computing deals with only one n-bit at a time.

For example, the state of a 3-qubit system is the superposition of eight basic states:

$$|\psi\rangle = \alpha_0|000\rangle + \alpha_1|001\rangle + \alpha_2|010\rangle + \alpha_3|011\rangle + \alpha_4|100\rangle + \alpha_5|101\rangle + \alpha_6|110\rangle + \alpha_7|111\rangle$$

An n-qubit can be written as a vector:

$$\begin{pmatrix} \alpha_0 \\ \alpha_1 \\ ... \\ \alpha_{2^n-1} \end{pmatrix} \in \mathbb{C}^{2^n}$$

The condition of normalisation must be verified using:

$$\sum_{i=0}^{2^n-1} |\alpha_i|^2 = 1$$

The measurement of an n-qubit with the norm 1 leads to a classical n-bit: $00...00$ with the probability $|\alpha_0|^2$, $00...01$ with the probability $|\alpha_1|^2$, ..., $11...11$ with the probability $|\alpha_{n^2-1}|^2$.

2.8.2 THE TOFFOLI GATE (CCNOT)

The Toffoli gate is an example of a gate that requires 3 qubits as input. If the state of the first two qubits is $|1\rangle$, then the gate exchanges $|0\rangle$ to $|1\rangle$ for the third qubit; otherwise, it keeps the third qubit unchanged.

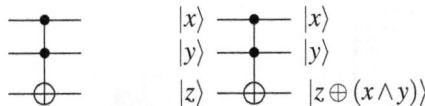

The Toffoli gate is a generalisation of the CNOT gate, also called CCNOT or CCX. If the first and the second inputs are both equal to 1, then the third qubit is reversed using the X gate:

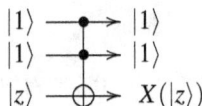

Otherwise, the third qubit remains unchanged:

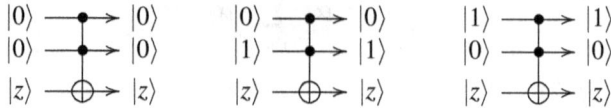

$$
\begin{array}{ccc}
|0\rangle \longrightarrow |0\rangle & |0\rangle \longrightarrow |0\rangle & |1\rangle \longrightarrow |1\rangle \\
|0\rangle \longrightarrow |0\rangle & |1\rangle \longrightarrow |1\rangle & |0\rangle \longrightarrow |0\rangle \\
|z\rangle \oplus |z\rangle & |z\rangle \oplus |z\rangle & |z\rangle \oplus |z\rangle
\end{array}
$$

The matrix M representing the Toffoli gate is an 8×8 matrix given as follows:

$$
M = \left(\begin{array}{cccccc|cc}
1 & 0 & 0 & 0 & 0 & 0 & 0 & 0 \\
0 & 1 & 0 & 0 & 0 & 0 & 0 & 0 \\
0 & 0 & 1 & 0 & 0 & 0 & 0 & 0 \\
0 & 0 & 0 & 1 & 0 & 0 & 0 & 0 \\
0 & 0 & 0 & 0 & 1 & 0 & 0 & 0 \\
0 & 0 & 0 & 0 & 0 & 1 & 0 & 0 \\
\hline
0 & 0 & 0 & 0 & 0 & 0 & 0 & 1 \\
0 & 0 & 0 & 0 & 0 & 0 & 1 & 0
\end{array}\right)
$$

We conclude this section with an important remark regarding quantum gates in general. A crucial characteristic, not present in the context of classical gates, lies in the fact that with quantum gates, it is possible to determine the inputs from the outputs.

2.8.3 H GATE: GENERAL FORMULA FOR $|00...0\rangle$

The general Hadamard gate (H gate), also called the Transformation of Hadamard, for a n-qubit $|\psi\rangle = |\psi_1 \psi_2 ... \psi_n\rangle$ is the application of the H gate for each 1-qubit $|\psi_i\rangle$ composing it. This transformation, $H^{\otimes n}$, is represented as:

$$
|\psi\rangle = |\psi_1 \psi_2 ... \psi_n\rangle \quad \xrightarrow{\;\;n\;\;} \boxed{H^{\otimes n}} \longrightarrow H^{\otimes n}(|\psi\rangle)
$$

Or, as:

$$
\begin{cases}
|\psi_1\rangle \longrightarrow \boxed{H} \longrightarrow \\
|\psi_2\rangle \longrightarrow \boxed{H} \longrightarrow \\
\;\vdots \quad \longrightarrow \quad \vdots \quad \longrightarrow \\
|\psi_n\rangle \longrightarrow \boxed{H} \longrightarrow
\end{cases} H^{\otimes n}(|\psi\rangle)
$$

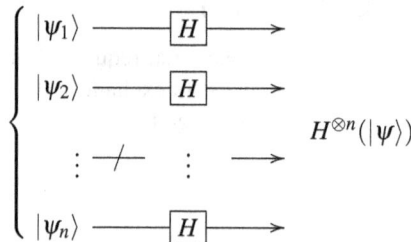

For $n = 2$, the Hadamard transformation of $|00\rangle$ is given in Section 2.7.5. For any value of n, consider the qubit $|0\rangle^{\otimes n} = |00...0\rangle$. It is known that $H(|0\rangle) = \frac{1}{\sqrt{2}}(|0\rangle + |1\rangle)$, then, $H^{\otimes n}(|0\rangle^{\otimes n})$ can be written as:

$$
H^{\otimes n}(|0\rangle^{\otimes n}) = H^{\otimes n}(|00...0\rangle) = H(|0\rangle)H(|0\rangle)...H(|0\rangle)
$$

$$H^{\otimes n}(|0\rangle^{\otimes n}) = \left(\frac{1}{\sqrt{2}}(|0\rangle + |1\rangle)\right) \otimes \left(\frac{1}{\sqrt{2}}(|0\rangle + |1\rangle)\right) \otimes \dots \otimes \left(\frac{1}{\sqrt{2}}(|0\rangle + |1\rangle)\right)$$

$$H^{\otimes n}(|0\rangle^{\otimes n}) = \frac{1}{\sqrt{2^n}}(|0\rangle + |1\rangle) \otimes (|0\rangle + |1\rangle) \otimes \dots \otimes (|0\rangle + |1\rangle)$$

$$H^{\otimes n}(|0\rangle^{\otimes n}) = \frac{1}{\sqrt{2^n}}(|00...0\rangle + |00...1\rangle + ... + |11...1\rangle)$$

This can be written as:

$$H^{\otimes n}(|0\rangle^{\otimes n}) = \frac{1}{\sqrt{2^n}} \sum_{x \in \{0,1\}^n} |x\rangle \tag{2.9}$$

For example, for $n = 3$, we can write:

$$H^{\otimes 3}(|000\rangle) = \frac{1}{\sqrt{2^3}}(|000\rangle + |001\rangle + |010\rangle + |011\rangle + |100\rangle + |101\rangle + |110\rangle + |111\rangle)$$

For any value of n, we can calculate the Hadamart transformation matrix $H^{\otimes n}$ as follows:

$$H^{\otimes n} = \frac{1}{\sqrt{2^n}} \left(\begin{array}{c|c} 1 \times H^{\otimes n-1} & 1 \times H^{\otimes n-1} \\ \hline 1 \times H^{\otimes n-1} & -1 \times H^{\otimes n-1} \end{array} \right)$$

For example, for $n = 1$, we have:

$$H^{\otimes 1} = \frac{1}{\sqrt{2}} \left(\begin{array}{cc} 1 & 1 \\ 1 & -1 \end{array} \right)$$

Then, for $n = 2$, we can calculate $H^{\otimes 2}$ as follows:

$$H^{\otimes 2} = \frac{1}{2} \left(\begin{array}{c|c} 1 \times H^{\otimes 1} & 1 \times H^{\otimes 1} \\ \hline 1 \times H^{\otimes 1} & -1 \times H^{\otimes 1} \end{array} \right) = \frac{1}{2} \left(\begin{array}{cc|cc} 1 & 1 & 1 & 1 \\ 1 & -1 & 1 & -1 \\ \hline 1 & 1 & -1 & -1 \\ 1 & -1 & -1 & 1 \end{array} \right)$$

2.9   ORACLE AND FUNCTION

2.9.1   ORACLE WITH ONE VARIABLE

An oracle is a quantum circuit associated with a function f. Here is what an oracle accomplishes for a function $f : \{0,1\} \mapsto \{0,1\}$. It gives us the action of the gate U_f on the basic qubits $|0\rangle$ and $|1\rangle$.

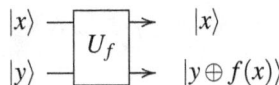

$$|x\rangle \longrightarrow \boxed{U_f} \longrightarrow |x\rangle$$
$$|y\rangle \longrightarrow \phantom{\boxed{U_f}} \longrightarrow |y \oplus f(x)\rangle$$

In the first line, as input, the oracle receives $|x\rangle$, and as output, it returns the same $|x\rangle$. In the second line, as input, the oracle receives the $|y\rangle$, but the output depends on the values of $|x\rangle$, $|y\rangle$, and the function f. This output can be $|0\rangle$ or $|1\rangle$ and it is given by the formula:

$$|y \oplus f(x)\rangle$$

In the case where $|y\rangle = |0\rangle$, the second output is equal to $|f(x)\rangle$. Then, consider this gate as a **function** gate. Its circuit is given as follows:

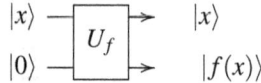

$$|x\rangle \;-\boxed{}\!\!\to\; |x\rangle$$
$$U_f$$
$$|0\rangle \;-\boxed{}\!\!\to\; |f(x)\rangle$$

As an example, consider the function f defined by $f(x) = x$. Then,

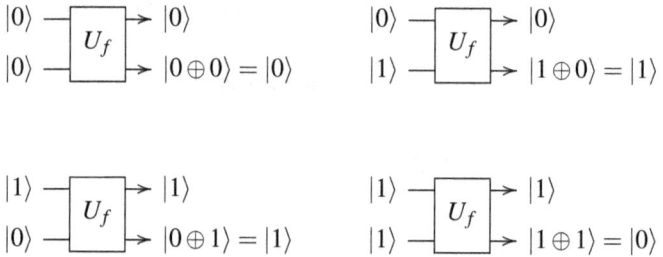

$$|0\rangle \;-\boxed{U_f}\!\!\to\; |0\rangle$$
$$|0\rangle \;-\boxed{U_f}\!\!\to\; |0 \oplus 0\rangle = |0\rangle$$

$$|0\rangle \;-\boxed{U_f}\!\!\to\; |0\rangle$$
$$|1\rangle \;-\boxed{U_f}\!\!\to\; |1 \oplus 0\rangle = |1\rangle$$

$$|1\rangle \;-\boxed{U_f}\!\!\to\; |1\rangle$$
$$|0\rangle \;-\boxed{U_f}\!\!\to\; |0 \oplus 1\rangle = |1\rangle$$

$$|1\rangle \;-\boxed{U_f}\!\!\to\; |1\rangle$$
$$|1\rangle \;-\boxed{U_f}\!\!\to\; |1 \oplus 1\rangle = |0\rangle$$

This oracle represents the CNOT gate. In other terms, a CNOT gate is an oracle associated with a function $f(x) = x$. A CNOT gate is equivalent to the classical Gate **XOR**.

2.9.2   ORACLE WITH TWO VARIABLES

In the same way as an oracle with one variable, an oracle with two variables is a quantum circuit associated with a function f, where $f : \{0,1\}^2 \mapsto \{0,1\}$. It is represented as follows:

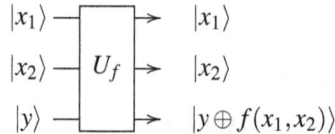

$$|x_1\rangle \;-\boxed{}\!\!\to\; |x_1\rangle$$
$$|x_2\rangle \;-\boxed{U_f}\!\!\to\; |x_2\rangle$$
$$|y\rangle \;-\boxed{}\!\!\to\; |y \oplus f(x_1,x_2)\rangle$$

In the first two lines, the oracle receives $|x_1\rangle$ and $|x_2\rangle$ as input and returns the same $|x_1\rangle$ and $|x_2\rangle$ as output. In the second line, the oracle receives the $|y\rangle$ and input, but the output depends on the values of $|x_1\rangle$, $|x_2\rangle$, $|y\rangle$, and the function f. This output can be $|0\rangle$ or $|1\rangle$ and it is given by the formula:

$$|y \oplus f(x_1,x_2)\rangle$$

As an example, consider the function $f(x_1, x_2) = 1$, if $x_1 = x_2 = 1$; otherwise, $f(x_1, x_2) = 0$. This oracle represents the Toffoli gate (see Section 2.8.2), also called the CCNOT gate, and it is equivalent to the classical Gate **AND**.

2.9.3 ORACLE WITH N VARIABLES

An oracle with n variables is a quantum circuit associated with a function f, where $f : \{0,1\}^n \mapsto \{0,1\}$. It is represented as follows:

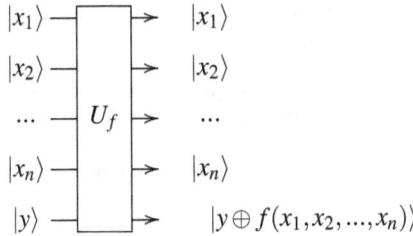

$$
\begin{array}{ccc}
|x_1\rangle & & |x_1\rangle \\
|x_2\rangle & & |x_2\rangle \\
\cdots & U_f & \cdots \\
|x_n\rangle & & |x_n\rangle \\
|y\rangle & & |y \oplus f(x_1, x_2, ..., x_n)\rangle
\end{array}
$$

2.9.4 ORACLE EXAMPLES

The CNOT and the CCNOT gates can be considered as Oracles. The following example shows another example of an Oracle, where $f(x_1, x_2, x_3) = x_1 \oplus x_2 \oplus x_3$.

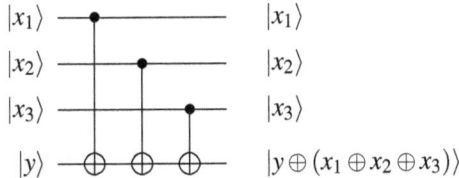

$$
\begin{array}{cc}
|x_1\rangle & |x_1\rangle \\
|x_2\rangle & |x_2\rangle \\
|x_3\rangle & |x_3\rangle \\
|y\rangle & |y \oplus (x_1 \oplus x_2 \oplus x_3)\rangle
\end{array}
$$

2.10 CONCLUSION

This chapter's objective is to provide readers with a solid understanding of the fundamental principles of quantum computing, from the behaviour of qubits on the Bloch sphere to the operations performed by quantum gates. It provided a foundational understanding of quantum programming, explaining its principles and implications. By investigating the fundamental concepts of quantum mechanics and demonstrating how they shape quantum computing, readers have gained insight into the unique computational paradigm offered by quantum technology. From the exploration of quantum gates to the challenges of error correction and fault tolerance, the chapter has highlighted the need for a quantum-native approach to programming. With the emergence of quantum programming languages such as Qiskit, Quipper, and Q#, the stage is set for advancements in computation. As the field continues to evolve, this chapter serves as a stepping stone for researchers and practitioners to harness the full potential of quantum algorithms and pave the way for a future of scalable and reliable quantum computation.

REFERENCES

1. Chuang, I.L., Gershenfeld, N., Kubinec, M.: Quantum computing with nuclear spins in liquid solution. Proceedings of the Royal Society of London. Series A: Mathematical, Physical and Engineering Sciences **454**(1969), 447–467 (1998).

2. Nakahara, M., Ohmi, T.: Quantum Computing: From Linear Algebra to Physical Realisations. CRC Press, UK (2008).

3. Deutsch, D.: Quantum theory, the church–turing principle and the universal quantum computer. Proceedings of the Royal Society of London. A. Mathematical and Physical Sciences **400**(1818), 97–117 (1985).

4. Feynman, R.P.: Simulating physics with computers. International Journal of Theoretical Physics **21**(6/7), 467–488 (1982).

5. Shor, P.W.: Algorithms for quantum computation: discrete logarithms and factoring. Proceedings 35th Annual Symposium on Foundations of Computer Science, 124–134 (1994).

6. Grover, L.K.: A fast quantum mechanical algorithm for database search. arXiv preprint quant-ph/9605043 (1996).

3 Fundamentals of Quantum Programming

Ahcene Bounceur, Mohammad Hammoudeh,
Bamidele Adebisi, Foudil Mir, and Madani Bezoui

3.1 INTRODUCTION

This chapter is dedicated to demystifying quantum programming. In contrast to classical programming, which functions using classical bits, quantum programming leverages the capabilities of qubits, which are the fundamental units of quantum computation. Qubits are derived from the principles of quantum mechanics and possess distinctive characteristics, including superposition and entanglement, that enable information representation and processing in ways that defy classical intuition.

The foundation of quantum programming lies in quantum gates, which are analogous to gates in classical logic. These gates, including the Pauli gates, such as the X, Y, and Z gates, the Hadamard gate, and the CNOT gate [2, 3], offer the capability to manipulate qubits and execute operations that serve as the foundation for quantum algorithms. Through the strategic coordination of quantum gate sequences, programmers are able to design quantum algorithms that exhibit superior efficiency in resolving complex problems when compared to classical algorithms.

Superdense coding is one of the remarkable algorithms that will be examined in this chapter. By taking advantage of quantum entanglement, this algorithm demonstrates the exceptional capabilities of quantum programming to transmit data with an unprecedented degree of efficiency. Superdense coding, which was invented in 1992 by Charles H. Bennett and Stephen J. Wiesner [3], enables a transmitter to encode and transmit two classical bits of information onto a single qubit. By employing quantum gates and entangled qubits, the receiver is capable of deciphering the information with exceptional precision, thereby exceeding the limits of classical communication efficiency. This chapter not only explains the conceptual underpinnings of quantum programming but also provides practical insights into its applications. Quantum programming grants access to uncharted domains of computation, including optimization, machine learning, and quantum simulations [4, 5].

DOI: 10.1201/9781003475286-3

3.2 QUANTUM PROGRAMMING

In this section, we use Qiskit library from IBM to program quantum circuits. This library enables the simulation of a quantum computer and the utilization of a real one. The programming language used is Python. We present two programs: the Hadamard gate and the X gate. The simulation of these circuits is done in two ways. The first one allows obtaining results in the form of quantum states (qubits).

While this simulation is not realistic, it serves as a reference to verify the accuracy of the programmed circuit. Alternatively, in the second option, the results are obtained based on the measured outputs and they are given in their probabilistic form. In this scenario, the measured outputs can be obtained either through simulation or by employing an actual quantum computer.

3.2.1 THE X GATE PROGRAM

In this section, we present the quantum program of the X gate circuit

$$—\boxed{X}—$$

in two ways: the quantum state output and the real probabilistic output.

3.2.1.1 Version 1: Quantum State Output

We use a *simulated* quantum computer to obtain the output quantum states of qubits. This program is given as follows:

```
1  import qiskit as q
2  simulator = q.Aer.get_backend('statevector_simulator')
3  circuit = q.QuantumCircuit(1)
4  circuit.initialize(0)
5  circuit.x(0)
6  print(circuit.draw(output='text'))
7  job = q.execute(circuit, simulator)
8  result = job.result()
9  output = result.get_statevector()
10 print(output)
```

Below, we go through the above program step by step with explanations and results.

- Line 1: The Python module to be imported beforehand is the Qiskit module, abbreviated by the single letter "q".
- Line 2: In this code, we do not use a real quantum computer; we use a local machine as a simulator with the option "statevector_simulator" to obtain the quantum states of qubits at the output of the circuit.
- Line 3: The instruction defines a quantum circuit named "circuit" and declares the number of quantum bits (here, 1) in the quantum register.

- Line 4: Sets (initializes) the value of the input to $|0\rangle$. It is possible to use the vector representation of $|0\rangle$ instead. As the quantum state $|0\rangle = 1|0\rangle + 0|1\rangle$ can be written in its vector version as $(1,0)$, this instruction can be written as `circuit.initialize([1,0])`.
- Line 5: Adds the X gate in the first line (0) of the quantum register.
- Line 6: To check the obtained circuit, this instruction allows to draw the circuit on the terminal. Figure 3.1 displays the resulting outcome.

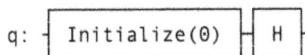

Figure 3.1   X gate circuit drawing.

- Line 7: Run the simulation.
- Line 8: Assign the obtained result to the variable `result`.
- Line 9: Assign the variable `output` the obtained state vector of the output.
- Line 10: Display in the terminal the coefficients α' and β' of the output: `[0.+0.j, 1.+0.j]`.
  In Python the letter j represents the letter i of the complex numbers. Thus, `0.+0.j` means 0+0i which is the real number 0 and `1.+0.j` means 1+0i which is the real number 1. Therefore, the output is the vector `[0,1]` which represents the vector representation of the quantum state $|1\rangle$.

Note that if the input is set to $|0\rangle$, then the initialization instruction of line 4 is not required and can be left out.

3.2.1.2   Version 2: Real Probabilistic Output

In this version, the quantum circuit is used to provide approximate probabilities instead of the quantum states of qubits. We start from the initial state $|0\rangle$, for which we apply an X gate to transform its quantum state to $|1\rangle$. A measurement results in a classical bit 1, with a probability of 1.

To better represent the reality of this circuit, it should be written in two lines. The first line corresponds to the qubit, denoted as q, and its transformation. The second line corresponds to the classical bit, denoted by c, which is used to store the measured outcome of the qubit.

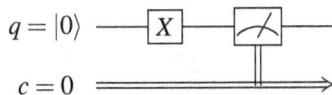

The program of the X gate circuit is given as follows, which is obtained by slightly modifying the previous program.

```
1  import qiskit as q
2  simulator = q.Aer.get_backend('qasm_simulator')
3  circuit = q.QuantumCircuit(1,1)
```

```
4   circuit.initialize(0)
5   circuit.x(0)
6   circuit.measure(0, 0)
7   print(circuit.draw(output='text'))
8   job = q.execute(circuit, simulator, shots=1000)
9   result = job.result()
10  output = result.get_counts(circuit)
11  print(output)
```

Below are the modified instructions:

- Line 2: Change the name of the simulator from "statevector_simulator" to "qasm_simulator".
- Line 3: `circuit = q.QuantumCircuit(1,1)` is replaced by, which means that there is one line in the quantum register like in the previous program. There is another line in the classical register which is used to obtain the measured qubit in the quantum register.
- Line 6: After adding the X gate in line 5, the measurement module is added using the instruction `circuit.measure(0, 0)`. The first 0 means the first line of the quantum register where the measurement module is added. The second 0 means the first line of the classical register, where the measured value of the output qubit is retrieved.
- Line 7: This instruction remains the same. However, the drawn circuit, illustrated in Figure 3.2, is different. The circuit is illustrated in Figure 3.2.

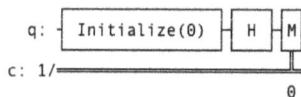

Figure 3.2 X gate circuit with the measurement module.

- Line 8: Add to the function `execute` the argument `shots=1000`. This means that the output is measured 1000 times.
- Line 10: Replace the instruction `output = result.get_statevector()` by the instruction `output = result.get_counts(circuit)`. In this case, there is no need to get the quantum state of the output, but the returned number of times is 0s and 1s. The obtained numbers are displayed in line 11.
- Line 11: Remains the same as in the previous program. However, the result is different. We obtain the following:
  `{'1': 1000}`.
  This means that the measured output is 1 with the probability of $|\alpha'|^2 = 1000/1000 = 1$ and 0 with the probability of $|\beta'|^2 = 0/1000 = 0$. Therefore, $\alpha' = 1$ and $\beta' = 0$. The output is then $0|0\rangle + 1|1\rangle = |1\rangle$.

3.2.2   THE HADAMARD GATE PROGRAM

This section gives the programming of the Hadamard Gate circuit

$$\boxed{H}$$

We start from the initial state $|0\rangle$, for which we apply a Hadamard gate to transform its quantum state to $\frac{1}{\sqrt{2}}|0\rangle + \frac{1}{\sqrt{2}}|1\rangle$. A measurement results in a classical bit 0 or 1, each with a probability of 0.5.

To better represent this circuit, it should be written in two lines. The first line corresponds to the qubit, denoted as q, and its transformation. The second line corresponds to the classical bit, denoted by c, which is used to store the measured outcome of the qubit.

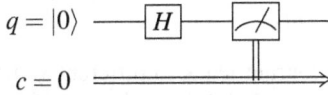

$$q = |0\rangle \;\; \boxed{H} \;\; \boxed{\angle}$$
$$c = 0 \;\; =========\Rightarrow$$

The program of the Hadamard gate circuit is given as follows:

```
1   import qiskit as q
2   simulator = q.Aer.get_backend('qasm_simulator')
3   circuit = q.QuantumCircuit(1,1)
4   circuit.initialize(0)
5   circuit.h(0)
6   circuit.measure(0, 0)
7   print(circuit.draw(output='text'))
8   job = q.execute(circuit, simulator, shots=1000)
9   result = job.result()
10  output = result.get_counts(circuit)
11  print(output)
```

Below is a step by step run-through of the above program with explanations and results.

- Line 1: The Python module to be imported beforehand is the Qiskit module, abbreviated by the single letter "q".
- Line 2: This code does not use a real quantum computer, instead, it uses a local machine as a simulator with the option "qasm_simulator".
- Line 3: The instruction defines a quantum circuit named "circuit" and declares the number of quantum bits (here, 1) followed by the number of classical bits (also 1, for measurement).
- Line 5: Add the H gate in line (0) of the quantum register.
- Line 6: Add the measurement module to the first line (0) of the quantum register and send the result to the first line (0) of the classical register.
- Line 7: Check the obtained circuit, this instruction allows to draw the circuit on the terminal. We obtain the result illustrated in Figure 3.3.

$$q: \;-\!\!\boxed{\texttt{Initialize(1,0)}}\!-\!\boxed{\text{H}}\!-\!\boxed{\text{M}}\!-$$

$$c: \; 1/\!=\!\!=\!\!=\!\!=\!\!=\!\!=\!\!=\!\!=\!\!=\!\!=\!\!=\!\!=\!\!=$$

$$0$$

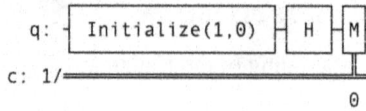

Figure 3.3 Qiskit Hadamard circuit drawing.

- Line 8: Run the simulation by specifying how many times the obtained qubit is measured (here is 1000).
- Line 9: Assign the obtained result to the variable `result`.
- Line 10: Assign to the variable `output` the number of measured 0s and 1s.
- Line 11: Display in the terminal the variable `output`. The result can be like: {'0': 498, '1': 502}.

Each execution of this program leads to different values of the number of "0" and "1". These values are the probabilities $|\alpha'|^2$ and $|\beta'|^2$. Each of them must be equal to $1/2 = 0.5$, which is the case here with $|\alpha'|^2 = 498/1000 = 0.498 \approx 0.5$ and $|\beta'|^2 = 502/1000 = 0.502 \approx 0.5$, i.e.,

$$|0\rangle \xrightarrow{\;H\;} \frac{1}{\sqrt{2}}|0\rangle + \frac{1}{\sqrt{2}}|1\rangle$$

To conclude, it is possible to use the library matplotlib to draw the histograms of the results. The previous program can be completed with the following lines.

```
1  import matplotlib.pyplot as plt
2  q.visualization.plot_histogram(counts)
3  plt.show()
```

The execution of these lines will draw Figure 3.4

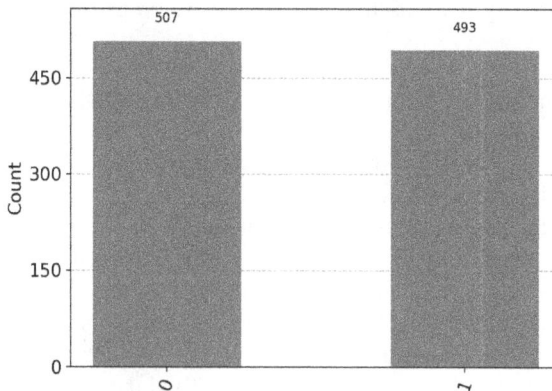

Figure 3.4 Qiskit Hadamard simulation results.

3.2.3   THE BELL STATE PROGRAM

In this section, we program the Bell State circuit

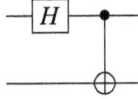

We start from the initial state $|0\rangle$, for which we apply a Hadamard gate to transform its quantum state to $\frac{1}{\sqrt{2}}|0\rangle + \frac{1}{\sqrt{2}}|1\rangle$. A measurement results in a classical bit 0 or 1, each with a probability of 0.5. Then, we apply the CNOT gate, which allows obtaining the Bell State or the Quantum Entanglement. This means that if the first output of the quantum register is $|0\rangle$, then the second output of the quantum register is $|0\rangle$ as well, and if the first output of the quantum register is $|1\rangle$, then the second output of the quantum register is also $|1\rangle$.

To better represent the reality of this circuit, it should be written in four lines. The first line corresponds to the first qubit, denoted as q_0, and its transformation. The second line corresponds to the second qubit, denoted as q_1, and its transformation. The third line corresponds to the first classical bit, denoted by c_0, which is used to store the measured outcome of the first qubit. The fourth line corresponds to the second classical bit, denoted by c_1, which is used to store the measured outcome of the second qubit.

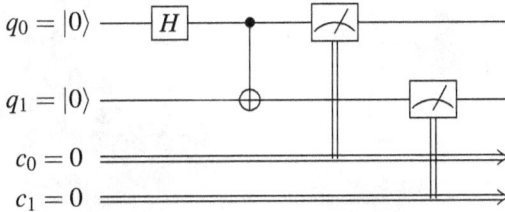

The program of the Hadamard gate circuit is given as follows:

```
1   import qiskit as q
2   simulator = q.Aer.get_backend('qasm_simulator')
3   circuit = q.QuantumCircuit(2,2)
4   circuit.initialize(0)
5   circuit.h(0)
6   circuit.cx(0,1)
7   circuit.measure([0,1], [0,1])
8   print(circuit.draw(output='text'))
9   job = q.execute(circuit, simulator, shots=1000)
10  result = job.result()
11  output = result.get_counts(circuit)
12  print(output)
13  import matplotlib.pyplot as plt
14  q.visualization.plot_histogram(output)
15  plt.show()
```

The execution of this program gives the following results:

- Line 8: Draw the following circuit:

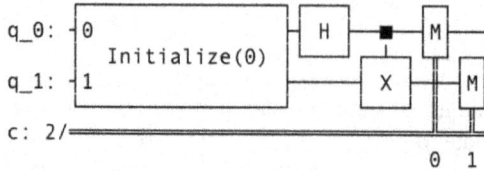

Figure 3.5   Bell circuit.

- Line 12: Display the following text:
  {'11': 514, '00': 486}
  Only the qubits $|00\rangle$ and $|11\rangle$ are generated in the output.
- Lines 13 to 15: The results in the form of histograms.

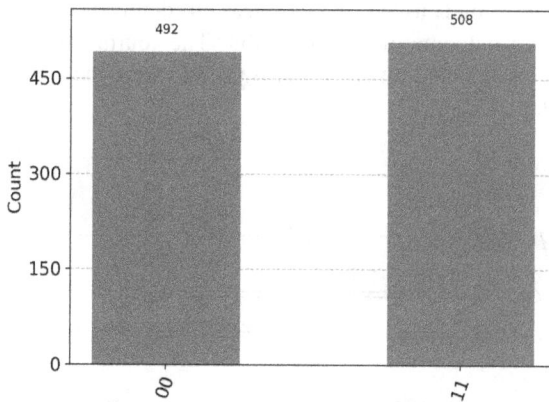

Figure 3.6   Bell State histogram.

3.2.4 THE SUPERDENSE CODING PROGRAM

The Superdense coding is a quantum protocol that enables two people to exchange information. The general schematic is given in Figure 3.7.

Alice wants to send a secure message to Bob. Alice can send four kinds of messages: 00, 01, 10, and 11. If Alice directly sends her message to Bob and suppose there is any interceptor (attacker) sniffing her message, then he can obtain a copy of it. To avoid this security weakness, the Superdense coding model proposes to add another person (Charlie) who sends the same qubit to Alice and Bob at the same time using the Bell State concept. Based on the message that Alice wants to send to Bob and on the qubit sent by Charlie, Alice makes some transformations, using gates, to

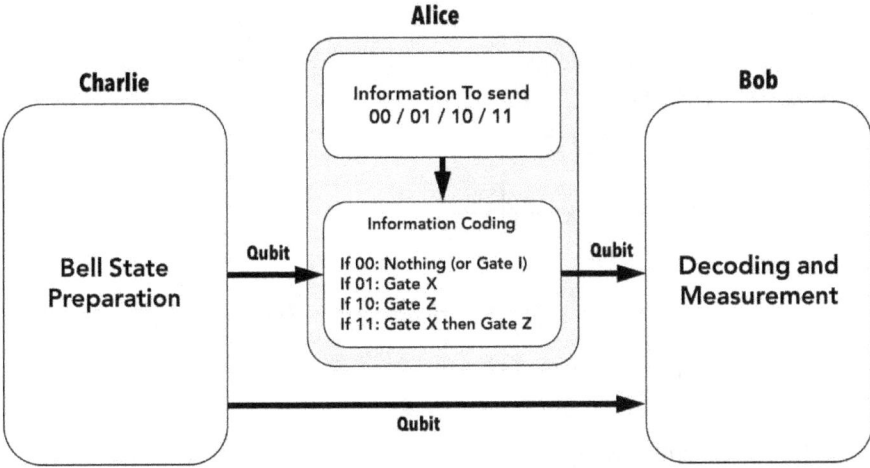

Figure 3.7   General schematic of the Superdense coding.

the qubit received from Charlie and sends it to Bob. The transformations that Alice applies on the qubit received by Charlie are based on the message she wants to send, and they are given as follows:

- Alice wants to send 00: she will not do any transformation to the qubit received from Charlie.
- Alice wants to send 01: she will apply Gate X to the qubit received from Charlie.
- Alice wants to send 10: she will apply Gate Z to the qubit received from Charlie.
- Alice wants to send 11: she will apply Gate X, then Gate Z to the qubit received from Charlie.

The Superdense coding concept can be formulated in another way. It can be considered a problem of finding which gates should be added in the first line between the Bell State circuit and the decoding circuit to obtain at the output the desired state for the input $|00\rangle$. This means that if the input state is $|00\rangle$, then, the output state should be $|ab\rangle$. This situation is illustrated as:

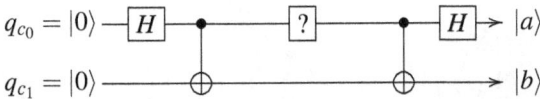

$$q_{c_0} = |0\rangle - \boxed{H} - \bullet - \boxed{?} - \bullet - \boxed{H} \rightarrow |a\rangle$$
$$q_{c_1} = |0\rangle - \oplus - \oplus \rightarrow |b\rangle$$

In the following, the symbol $\xrightarrow{G^1}$ represents the application of the gate G only to line 1. It is the same as GI, which means for the first line we apply the gate G and for the second line we apply the gate I (identity).

Case of $|00\rangle$

In the case of the input state $|00\rangle$, we obtain in the input $|00\rangle$ without adding any gates or the identity gate.

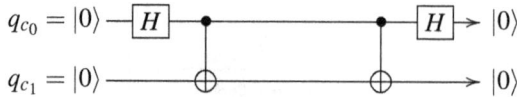

or

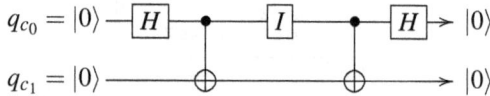

This can be explained by:

$$|00\rangle \xrightarrow{Bell} \frac{1}{\sqrt{2}}|00\rangle + \frac{1}{\sqrt{2}}|11\rangle \xrightarrow{I^1=II} \frac{1}{\sqrt{2}}|00\rangle + \frac{1}{\sqrt{2}}|11\rangle \xrightarrow{CNOT}$$

$$\frac{1}{\sqrt{2}}|00\rangle + \frac{1}{\sqrt{2}}|10\rangle = \left(\frac{1}{\sqrt{2}}|0\rangle + \frac{1}{\sqrt{2}}|1\rangle\right) \otimes |0\rangle =$$

$$|+\rangle \otimes |0\rangle = |+0\rangle \xrightarrow{H^1=HI} H(|+\rangle) \otimes |0\rangle = |00\rangle$$

In terms of vector, we obtain:

$$\begin{pmatrix} 1 \\ 0 \\ 0 \\ 0 \end{pmatrix} \xrightarrow{Bell} \frac{1}{\sqrt{2}} \begin{pmatrix} 1 \\ 0 \\ 0 \\ 1 \end{pmatrix} \xrightarrow{I^1=II} \frac{1}{\sqrt{2}} \begin{pmatrix} 1 \\ 0 \\ 0 \\ 1 \end{pmatrix} \xrightarrow{CNOT}$$

$$\frac{1}{\sqrt{2}} \begin{pmatrix} 1 \\ 0 \\ 1 \\ 0 \end{pmatrix} \xrightarrow{H^1=HI} \frac{1}{\sqrt{2}} H^1 \begin{pmatrix} 1 \\ 0 \\ 1 \\ 0 \end{pmatrix} = \frac{1}{\sqrt{2}} H^1 \left[\begin{pmatrix} 1 \\ 0 \\ 0 \\ 0 \end{pmatrix} + \begin{pmatrix} 0 \\ 0 \\ 1 \\ 0 \end{pmatrix} \right] =$$

$$\frac{1}{\sqrt{2}} H^1 \left[\begin{pmatrix} 1 \\ 0 \end{pmatrix} \otimes \begin{pmatrix} 1 \\ 0 \end{pmatrix} + \begin{pmatrix} 0 \\ 1 \end{pmatrix} \otimes \begin{pmatrix} 1 \\ 0 \end{pmatrix} \right] = \frac{1}{\sqrt{2}} \left[H \begin{pmatrix} 1 \\ 0 \end{pmatrix} \otimes \begin{pmatrix} 1 \\ 0 \end{pmatrix} + H \begin{pmatrix} 0 \\ 1 \end{pmatrix} \otimes \begin{pmatrix} 1 \\ 0 \end{pmatrix} \right]$$

$$\frac{1}{\sqrt{2}} \left[\frac{1}{\sqrt{2}} \begin{pmatrix} 1 \\ 1 \end{pmatrix} \otimes \begin{pmatrix} 1 \\ 0 \end{pmatrix} + \frac{1}{\sqrt{2}} \begin{pmatrix} 1 \\ -1 \end{pmatrix} \otimes \begin{pmatrix} 1 \\ 0 \end{pmatrix} \right] =$$

$$\frac{1}{2}\left[\begin{pmatrix}1\\0\\1\\0\end{pmatrix}+\begin{pmatrix}1\\0\\-1\\0\end{pmatrix}\right]=\frac{1}{2}\begin{pmatrix}2\\0\\0\\0\end{pmatrix}=\begin{pmatrix}1\\0\\0\\0\end{pmatrix}=|00\rangle$$

This can be calculated using directly the transformation HI, which is given by:

$$HI=\frac{1}{\sqrt{2}}\begin{pmatrix}1&1\\1&-1\end{pmatrix}\otimes\begin{pmatrix}1&0\\0&1\end{pmatrix}=\frac{1}{\sqrt{2}}\begin{pmatrix}1&0&1&0\\0&1&0&1\\1&0&-1&0\\0&1&0&-1\end{pmatrix}\qquad(3.1)$$

$$\frac{1}{\sqrt{2}}\begin{pmatrix}1\\0\\1\\0\end{pmatrix}\xrightarrow{HI}\frac{1}{2}\begin{pmatrix}1&0&1&0\\0&1&0&1\\1&0&-1&0\\0&1&0&-1\end{pmatrix}\begin{pmatrix}1\\0\\1\\0\end{pmatrix}=\frac{1}{2}\begin{pmatrix}2\\0\\0\\0\end{pmatrix}=\begin{pmatrix}1\\0\\0\\0\end{pmatrix}=|00\rangle$$

Case of $|01\rangle$

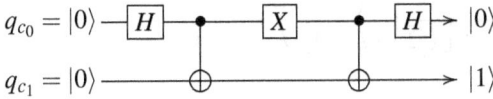

This can be explained by:

$$|00\rangle\xrightarrow{Bell}\frac{1}{\sqrt{2}}|00\rangle+\frac{1}{\sqrt{2}}|11\rangle\xrightarrow{X^1}\frac{1}{\sqrt{2}}|10\rangle+\frac{1}{\sqrt{2}}|01\rangle\xrightarrow{CNOT}$$

$$\frac{1}{\sqrt{2}}|11\rangle+\frac{1}{\sqrt{2}}|01\rangle=\left(\frac{1}{\sqrt{2}}|1\rangle+\frac{1}{\sqrt{2}}|0\rangle\right)\otimes|1\rangle=$$

$$|+\rangle\otimes|1\rangle=|+1\rangle\xrightarrow{H^1}H(|+\rangle)\otimes|1\rangle=|01\rangle$$

In terms of vectors, we obtain:

$$\begin{pmatrix}1\\0\\0\\0\end{pmatrix}\xrightarrow{Bell}\frac{1}{\sqrt{2}}\begin{pmatrix}1\\0\\0\\1\end{pmatrix}\xrightarrow{X^1}\frac{1}{\sqrt{2}}X^1\begin{pmatrix}1\\0\\0\\1\end{pmatrix}=$$

$$\frac{1}{\sqrt{2}}X^1\left[\begin{pmatrix}1\\0\\0\\0\end{pmatrix}+\begin{pmatrix}0\\0\\0\\1\end{pmatrix}\right]=\frac{1}{\sqrt{2}}X^1\left[\begin{pmatrix}1\\0\end{pmatrix}\otimes\begin{pmatrix}1\\0\end{pmatrix}+\begin{pmatrix}0\\1\end{pmatrix}\otimes\begin{pmatrix}0\\1\end{pmatrix}\right]=$$

$$\frac{1}{\sqrt{2}}\left[X\begin{pmatrix}1\\0\end{pmatrix}\otimes\begin{pmatrix}1\\0\end{pmatrix}+X\begin{pmatrix}0\\1\end{pmatrix}\otimes\begin{pmatrix}0\\1\end{pmatrix}\right]=\frac{1}{\sqrt{2}}\left[\begin{pmatrix}0\\1\end{pmatrix}\otimes\begin{pmatrix}1\\0\end{pmatrix}+\begin{pmatrix}1\\0\end{pmatrix}\otimes\begin{pmatrix}0\\1\end{pmatrix}\right]=$$

$$\frac{1}{\sqrt{2}}\left[\begin{pmatrix}0\\0\\1\\0\end{pmatrix}+\begin{pmatrix}0\\1\\0\\0\end{pmatrix}\right]=\frac{1}{\sqrt{2}}\begin{pmatrix}0\\1\\1\\0\end{pmatrix}\xrightarrow{CNOT}$$

$$\frac{1}{\sqrt{2}}\begin{pmatrix}0\\1\\0\\1\end{pmatrix}\xrightarrow{H^1}\frac{1}{\sqrt{2}}H^1\begin{pmatrix}0\\1\\0\\1\end{pmatrix}=\frac{1}{\sqrt{2}}H^1\left[\begin{pmatrix}0\\1\\0\\0\end{pmatrix}+\begin{pmatrix}0\\0\\0\\1\end{pmatrix}\right]=$$

$$\frac{1}{\sqrt{2}}H^1\left[\begin{pmatrix}1\\0\end{pmatrix}\otimes\begin{pmatrix}0\\1\end{pmatrix}+\begin{pmatrix}0\\1\end{pmatrix}\otimes\begin{pmatrix}0\\1\end{pmatrix}\right]=\frac{1}{\sqrt{2}}\left[H\begin{pmatrix}1\\0\end{pmatrix}\otimes\begin{pmatrix}0\\1\end{pmatrix}+H\begin{pmatrix}0\\1\end{pmatrix}\otimes\begin{pmatrix}0\\1\end{pmatrix}\right]=$$

$$\frac{1}{\sqrt{2}}\left[\frac{1}{\sqrt{2}}\begin{pmatrix}1\\1\end{pmatrix}\otimes\begin{pmatrix}0\\1\end{pmatrix}+\frac{1}{\sqrt{2}}\begin{pmatrix}1\\-1\end{pmatrix}\otimes\begin{pmatrix}0\\1\end{pmatrix}\right]=$$

$$\frac{1}{2}\left[\begin{pmatrix}0\\1\\0\\1\end{pmatrix}+\begin{pmatrix}0\\1\\0\\-1\end{pmatrix}\right]=\frac{1}{2}\begin{pmatrix}0\\2\\0\\0\end{pmatrix}=\begin{pmatrix}0\\1\\0\\0\end{pmatrix}=|01\rangle$$

This can be calculated using directly the transformation HI, given above by Equation (3.1).

$$\frac{1}{\sqrt{2}}\begin{pmatrix}0\\1\\0\\1\end{pmatrix}\xrightarrow{HI}\frac{1}{2}\begin{pmatrix}1&0&1&0\\0&1&0&1\\1&0&-1&0\\0&1&0&-1\end{pmatrix}\begin{pmatrix}0\\1\\0\\1\end{pmatrix}=\frac{1}{2}\begin{pmatrix}0\\2\\0\\0\end{pmatrix}=\begin{pmatrix}0\\1\\0\\0\end{pmatrix}=|01\rangle$$

Case of $|10\rangle$

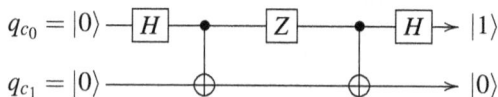

This can be explained by:

$$|00\rangle\xrightarrow{Bell}\frac{1}{\sqrt{2}}|00\rangle+\frac{1}{\sqrt{2}}|11\rangle\xrightarrow{Z^1}\frac{1}{\sqrt{2}}|00\rangle-\frac{1}{\sqrt{2}}|11\rangle\xrightarrow{CNOT}$$

$$\frac{1}{\sqrt{2}}|00\rangle - \frac{1}{\sqrt{2}}|10\rangle = \left(\frac{1}{\sqrt{2}}|0\rangle - \frac{1}{\sqrt{2}}|1\rangle\right) \otimes |0\rangle =$$

$$|-\rangle \otimes |0\rangle = |-0\rangle \xrightarrow{H^1} H(|-\rangle) \otimes |0\rangle = |10\rangle$$

In terms of vectors, we obtain:

$$\begin{pmatrix} 1 \\ 0 \\ 0 \\ 0 \end{pmatrix} \xrightarrow{Bell} \frac{1}{\sqrt{2}} \begin{pmatrix} 1 \\ 0 \\ 0 \\ 1 \end{pmatrix} \xrightarrow{Z^1} \frac{1}{\sqrt{2}} Z^1 \begin{pmatrix} 1 \\ 0 \\ 0 \\ 1 \end{pmatrix} =$$

$$\frac{1}{\sqrt{2}} Z^1 \left[\begin{pmatrix} 1 \\ 0 \\ 0 \\ 0 \end{pmatrix} + \begin{pmatrix} 0 \\ 0 \\ 0 \\ 1 \end{pmatrix} \right] = \frac{1}{\sqrt{2}} Z^1 \left[\begin{pmatrix} 1 \\ 0 \end{pmatrix} \otimes \begin{pmatrix} 1 \\ 0 \end{pmatrix} + \begin{pmatrix} 0 \\ 1 \end{pmatrix} \otimes \begin{pmatrix} 0 \\ 1 \end{pmatrix} \right] =$$

$$\frac{1}{\sqrt{2}} \left[Z\begin{pmatrix} 1 \\ 0 \end{pmatrix} \otimes \begin{pmatrix} 1 \\ 0 \end{pmatrix} + Z\begin{pmatrix} 0 \\ 1 \end{pmatrix} \otimes \begin{pmatrix} 0 \\ 1 \end{pmatrix} \right] = \frac{1}{\sqrt{2}} \left[\begin{pmatrix} 1 \\ 0 \end{pmatrix} \otimes \begin{pmatrix} 1 \\ 0 \end{pmatrix} + \begin{pmatrix} 0 \\ -1 \end{pmatrix} \otimes \begin{pmatrix} 0 \\ 1 \end{pmatrix} \right] =$$

$$\frac{1}{\sqrt{2}} \left[\begin{pmatrix} 1 \\ 0 \\ 0 \\ 0 \end{pmatrix} + \begin{pmatrix} 0 \\ 0 \\ 0 \\ -1 \end{pmatrix} \right] = \frac{1}{\sqrt{2}} \begin{pmatrix} 1 \\ 0 \\ 0 \\ -1 \end{pmatrix} \xrightarrow{CNOT}$$

$$\frac{1}{\sqrt{2}} \begin{pmatrix} 1 \\ 0 \\ -1 \\ 0 \end{pmatrix} \xrightarrow{H^1} \frac{1}{\sqrt{2}} H^1 \begin{pmatrix} 1 \\ 0 \\ -1 \\ 0 \end{pmatrix} = \frac{1}{\sqrt{2}} H^1 \left[\begin{pmatrix} 1 \\ 0 \\ 0 \\ 0 \end{pmatrix} - \begin{pmatrix} 0 \\ 0 \\ 1 \\ 0 \end{pmatrix} \right] =$$

$$\frac{1}{\sqrt{2}} H^1 \left[\begin{pmatrix} 1 \\ 0 \end{pmatrix} \otimes \begin{pmatrix} 1 \\ 0 \end{pmatrix} - \begin{pmatrix} 0 \\ 1 \end{pmatrix} \otimes \begin{pmatrix} 1 \\ 0 \end{pmatrix} \right] = \frac{1}{\sqrt{2}} \left[H\begin{pmatrix} 1 \\ 0 \end{pmatrix} \otimes \begin{pmatrix} 1 \\ 0 \end{pmatrix} - H\begin{pmatrix} 0 \\ 1 \end{pmatrix} \otimes \begin{pmatrix} 1 \\ 0 \end{pmatrix} \right] =$$

$$\frac{1}{\sqrt{2}} \left[\frac{1}{\sqrt{2}} \begin{pmatrix} 1 \\ 1 \end{pmatrix} \otimes \begin{pmatrix} 1 \\ 0 \end{pmatrix} - \frac{1}{\sqrt{2}} \begin{pmatrix} 1 \\ -1 \end{pmatrix} \otimes \begin{pmatrix} 1 \\ 0 \end{pmatrix} \right] =$$

$$\frac{1}{2} \left[\begin{pmatrix} 1 \\ 0 \\ 1 \\ 0 \end{pmatrix} - \begin{pmatrix} 1 \\ 0 \\ -1 \\ 0 \end{pmatrix} \right] = \frac{1}{2} \left[\begin{pmatrix} 1 \\ 0 \\ 1 \\ 0 \end{pmatrix} + \begin{pmatrix} -1 \\ 0 \\ 1 \\ 0 \end{pmatrix} \right] = \frac{1}{2} \begin{pmatrix} 0 \\ 0 \\ 2 \\ 0 \end{pmatrix} = \begin{pmatrix} 0 \\ 0 \\ 1 \\ 0 \end{pmatrix} = |10\rangle$$

This can be calculated using directly the transformation HI, given above by Equation (3.1).

$$\frac{1}{\sqrt{2}}\begin{pmatrix} 1 \\ 0 \\ -1 \\ 0 \end{pmatrix} \xmapsto{HI} \frac{1}{2}\begin{pmatrix} 1 & 0 & 1 & 0 \\ 0 & 1 & 0 & 1 \\ 1 & 0 & -1 & 0 \\ 0 & 1 & 0 & -1 \end{pmatrix}\begin{pmatrix} 1 \\ 0 \\ -1 \\ 0 \end{pmatrix} = \frac{1}{2}\begin{pmatrix} 0 \\ 0 \\ 2 \\ 0 \end{pmatrix} = \begin{pmatrix} 0 \\ 0 \\ 1 \\ 0 \end{pmatrix} = |10\rangle$$

Case of $|11\rangle$

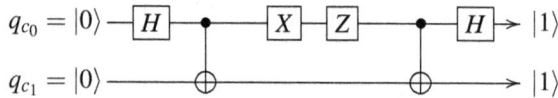

$$
\begin{array}{l}
q_{c_0} = |0\rangle \;-\boxed{H}\!-\!\bullet\!-\!-\boxed{X}\!-\!\boxed{Z}\!-\!-\!\bullet\!-\!-\boxed{H}\!\rightarrow\; |1\rangle \\
q_{c_1} = |0\rangle \;-\!-\!-\!-\!-\!\oplus\!-\!-\!-\!-\!-\!-\oplus\!-\!-\!-\!\rightarrow\; |1\rangle
\end{array}
$$

This can be explained by:

$$|00\rangle \xmapsto{Bell} \frac{1}{\sqrt{2}}|00\rangle + \frac{1}{\sqrt{2}}|11\rangle \xmapsto{X^1} \frac{1}{\sqrt{2}}|10\rangle + \frac{1}{\sqrt{2}}|01\rangle \xmapsto{Z^1}$$

$$\frac{1}{\sqrt{2}} - |1\rangle \otimes |0\rangle + \frac{1}{\sqrt{2}}|0\rangle \otimes |1\rangle \xmapsto{CNOT}$$

$$\frac{1}{\sqrt{2}} - |1\rangle \otimes |1\rangle + \frac{1}{\sqrt{2}}|0\rangle \otimes |1\rangle = \left(\frac{-1}{\sqrt{2}}|1\rangle + \frac{1}{\sqrt{2}}|0\rangle\right) \otimes |1\rangle =$$

$$|-\rangle \otimes |1\rangle = |-1\rangle \xmapsto{H^1} H(|-\rangle) \otimes |1\rangle = |11\rangle$$

In terms of vector, we obtain:

$$\begin{pmatrix} 1 \\ 0 \\ 0 \\ 0 \end{pmatrix} \xmapsto{Bell} \frac{1}{\sqrt{2}}\begin{pmatrix} 1 \\ 0 \\ 0 \\ 1 \end{pmatrix} \xmapsto{X^1} \frac{1}{\sqrt{2}}X^1\begin{pmatrix} 1 \\ 0 \\ 0 \\ 1 \end{pmatrix} =$$

$$\frac{1}{\sqrt{2}}X^1\left[\begin{pmatrix} 1 \\ 0 \\ 0 \\ 0 \end{pmatrix} + \begin{pmatrix} 0 \\ 0 \\ 0 \\ 1 \end{pmatrix}\right] = \frac{1}{\sqrt{2}}X^1\left[\begin{pmatrix} 1 \\ 0 \end{pmatrix} \otimes \begin{pmatrix} 1 \\ 0 \end{pmatrix} + \begin{pmatrix} 0 \\ 1 \end{pmatrix} \otimes \begin{pmatrix} 0 \\ 1 \end{pmatrix}\right] =$$

$$\frac{1}{\sqrt{2}}\left[X\begin{pmatrix} 1 \\ 0 \end{pmatrix} \otimes \begin{pmatrix} 1 \\ 0 \end{pmatrix} + X\begin{pmatrix} 0 \\ 1 \end{pmatrix} \otimes \begin{pmatrix} 0 \\ 1 \end{pmatrix}\right] = \frac{1}{\sqrt{2}}\left[\begin{pmatrix} 0 \\ 1 \end{pmatrix} \otimes \begin{pmatrix} 1 \\ 0 \end{pmatrix} + \begin{pmatrix} 1 \\ 0 \end{pmatrix} \otimes \begin{pmatrix} 0 \\ 1 \end{pmatrix}\right] \xmapsto{Z^1}$$

$$\frac{1}{\sqrt{2}}\left[Z\binom{0}{1}\otimes\binom{1}{0}+Z\binom{1}{0}\otimes\binom{0}{1}\right]=\frac{1}{\sqrt{2}}\left[\binom{0}{-1}\otimes\binom{1}{0}+\binom{1}{0}\otimes\binom{0}{1}\right]=$$

$$\frac{1}{\sqrt{2}}\left[\begin{pmatrix}0\\0\\-1\\0\end{pmatrix}+\begin{pmatrix}0\\1\\0\\0\end{pmatrix}\right]=\frac{1}{\sqrt{2}}\begin{pmatrix}0\\1\\-1\\0\end{pmatrix}\xrightarrow{CNOT}$$

$$\frac{1}{\sqrt{2}}\begin{pmatrix}0\\1\\0\\-1\end{pmatrix}\xrightarrow{H^1}\frac{1}{\sqrt{2}}H^1\begin{pmatrix}0\\1\\0\\-1\end{pmatrix}=\frac{1}{\sqrt{2}}H^1\left[\begin{pmatrix}0\\1\\0\\0\end{pmatrix}-\begin{pmatrix}0\\0\\0\\1\end{pmatrix}\right]=$$

$$\frac{1}{\sqrt{2}}H^1\left[\binom{1}{0}\otimes\binom{0}{1}-\binom{0}{1}\otimes\binom{0}{1}\right]=\frac{1}{\sqrt{2}}\left[H\binom{1}{0}\otimes\binom{0}{1}-H\binom{0}{1}\otimes\binom{0}{1}\right]=$$

$$\frac{1}{\sqrt{2}}\left[\frac{1}{\sqrt{2}}\binom{1}{1}\otimes\binom{0}{1}-\frac{1}{\sqrt{2}}\binom{1}{-1}\otimes\binom{0}{1}\right]=$$

$$\frac{1}{2}\left[\begin{pmatrix}0\\1\\0\\1\end{pmatrix}+\begin{pmatrix}0\\-1\\0\\1\end{pmatrix}\right]=\frac{1}{2}\begin{pmatrix}0\\0\\0\\2\end{pmatrix}=\begin{pmatrix}0\\0\\0\\1\end{pmatrix}=|11\rangle$$

This can be calculated using directly the transformation HI, given above by Equation (3.1).

$$\frac{1}{\sqrt{2}}\begin{pmatrix}0\\1\\0\\-1\end{pmatrix}\xrightarrow{HI}\frac{1}{2}\begin{pmatrix}1&0&1&0\\0&1&0&1\\1&0&-1&0\\0&1&0&-1\end{pmatrix}\begin{pmatrix}0\\1\\0\\-1\end{pmatrix}=\frac{1}{2}\begin{pmatrix}0\\0\\0\\2\end{pmatrix}=\begin{pmatrix}0\\0\\0\\1\end{pmatrix}=|11\rangle$$

3.2.4.1 All Cases in the Same Circuit

The quantum circuit of the Superdense coding concept is given as follows:

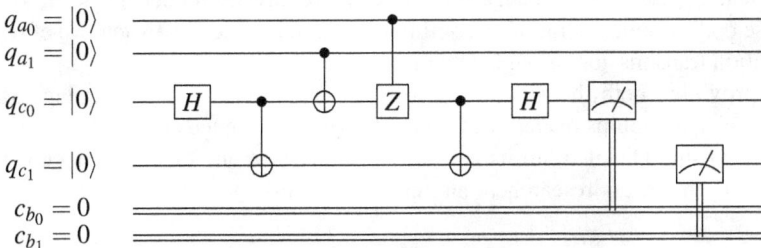

The corresponding program for the Superdense coding program is given as follows:

```
import qiskit as q
simulator = q.Aer.get_backend('qasm_simulator')
circuit = q.QuantumCircuit(4,2)
circuit.h(2)
circuit.cx(2,3)
circuit.cx(1,2)
circuit.cz(0,2)
circuit.cx(2,3)
circuit.h(2)
circuit.measure([2,3], [0,1])
print(circuit.draw(output='text'))
job = q.execute(circuit, simulator, shots=1000)
result = job.result()
output = result.get_counts(circuit)
print(output)
```

As you can see, the initialization part is removed from the program since implicitly the initialization is set to 0.

Finally, Alice transmits information composed of two bits to Bob, but she sends him only a single qubit (even though Bob receives two qubits in total: one from Alice and one from Charlie). Furthermore, it is a secure transmission protocol. Indeed, if Eve intercepts the qubit that Alice sends to Bob, she cannot extract any information from it because this qubit is in the form $\frac{1}{\sqrt{2}}(\pm|0\rangle \pm |1\rangle)$. Therefore, its measurement yields either 0 or 1, and it does not allow Eve to conclude anything about the information Alice intended to transmit.

3.3 CONCLUSION

This chapter provides an introduction to Qiskit, an open-source software framework designed by IBM to facilitate quantum programming. Through the strategic coordination of quantum gates, programmers can harness the power of quantum computation to tackle optimization, machine learning, and quantum simulation tasks.

The chapter not only explains the conceptual underpinnings of quantum programming but also offers practical insights into its applications, exemplified through the implementation of fundamental quantum gates like the Hadamard gate and the X gate using Qiskit. Additionally, it explores more advanced concepts such as Superdense coding, demonstrating the ability of quantum protocols to achieve efficient information transmission through gate manipulations.

By providing both theoretical explanations and hands-on programming examples, this chapter equips readers with the necessary knowledge and tools to program algorithms using Qiskit. With its user-friendly interface and extensive documentation, Qiskit empowers researchers and enthusiasts alike to explore the frontiers of quantum programming, driving innovation and discovery in this rapidly emerging field.

REFERENCES

1. LaPierre, R.: Quantum Gates, pp. 101–125. Springer, Cham (2021). https://doi.org/10.1007/978-3-030-69318-3_7

2. Hughes, C., Isaacson, J., Perry, A., Sun, R.F., Turner, J.: Quantum Gates, pp. 49–57. Springer, Cham (2021). https://doi.org/10.1007/978-3-030-61601-4_6

3. Bennett, C.H., Wiesner, S.J.: Communication via one- and two-particle operators on einstein-podolsky-rosen states. Physical Review Letters **69** (20), 2881–2884 (1992).

4. Terhal, B.M.: In: Kao, M.-Y. (ed.): Quantum Dense Coding, pp. 703–705. Springer, Boston, MA (2008). https://doi.org/10.1007/978-0-387-30162-4_314

5. Terhal, B.M.: In: Kao, M.-Y. (ed.): Quantum Dense Coding, pp. 1695–1698. Springer, New York, NY (2016). https://doi.org/10.1007/978-1-4939-2864-4_314

4 Seminal Quantum Algorithms Description and Programming

Ahcene Bounceur, Mohammad Hammoudeh,
Bamidele Adebisi, and Mostefa Kara

4.1 INTRODUCTION

Quantum computing is characterised by fundamentally distinct computational methodologies, which require a profound understanding of algorithmic and programming foundational principles. Central to this discipline is the intricate interplay of qubits, the fundamental units of quantum information processing, and the quantum gates responsible for their manipulation. This understanding helps in the exploration of quantum algorithms, where conventional paradigms of computation undergo scrutiny and revision.

This chapter serves as a reference for individuals already versed in quantum computation [1, 2], aiming to expand their practical proficiency in algorithm implementation through Qiskit [3], IBM's quantum computing library. Our exploration starts with a thorough familiarisation of the Qiskit framework, explaining its functionalities pertaining to the development and validation of quantum algorithms. Following this introduction to Qiskit, we revisit the concept of the oracle, an essential component in quantum algorithms, alongside its variations, such as the phase oracle. This introduction lays the groundwork, paving the way for the detailed exploration of prominent quantum algorithms.

Qiskit, developed by IBM, is a leading platform to democratise quantum computing, providing a comprehensive and open-source platform for quantum development. Known for its accessible interface and robust functionalities, Qiskit facilitates the engagement of researchers, developers, and enthusiasts in the domain of quantum information science.

This chapter gives an introduction to aid in understanding the fundamental concepts and practical applications of Qiskit. It provides hands-on experience in leveraging Qiskit to implement quantum algorithms, simulate quantum circuits, and interact with quantum processors. The main contribution of this chapter resides in the exposition of fundamental quantum algorithms, complemented by their respective

DOI: 10.1201/9781003475286-4

demonstrations and implementations using Qiskit. Specifically, the chapter presents three algorithms: Deutsch, Deutsch–Jozsa, and Bernstein–Vazirani. Each exposition of these algorithms provides profound insights into the capabilities and prospects of quantum computation.

4.2 DESIGNING FUNCTION CIRCUITS

This section shows how to design some circuits for a given function.

4.2.1 CIRCUIT 1: AND FUNCTION

Circuit 1 allows to design the function $f(x)$ such that $f(x) = 0$ for all $x \neq x_0$ and $f(x_0) = 1$. Consider a case with three qubits and $x_0 = |111\rangle$. This case can be obtained using the Toffoli gate represented by the circuit of Figure 4.1. This circuit is equivalent to the classical gate AND, where the output is equal to 0 only when all the inputs are equal to 1.

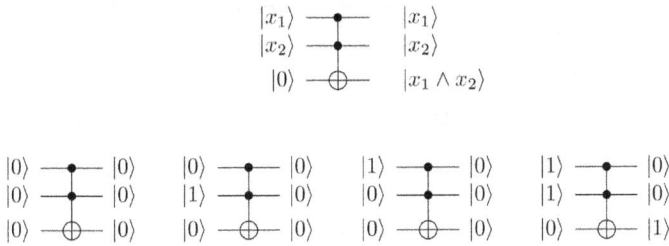

$$
\begin{array}{ll}
|x_1\rangle & |x_1\rangle \\
|x_2\rangle & |x_2\rangle \\
|0\rangle & |x_1 \wedge x_2\rangle
\end{array}
$$

$$
\begin{array}{llll}
|0\rangle \quad |0\rangle & |0\rangle \quad |0\rangle & |1\rangle \quad |0\rangle & |1\rangle \quad |0\rangle \\
|0\rangle \quad |0\rangle & |1\rangle \quad |0\rangle & |0\rangle \quad |0\rangle & |1\rangle \quad |0\rangle \\
|0\rangle \quad |0\rangle & |0\rangle \quad |0\rangle & |0\rangle \quad |0\rangle & |0\rangle \quad |1\rangle
\end{array}
$$

Figure 4.1 Quantum AND gate.

Consider the case of the AND gate with three inputs.

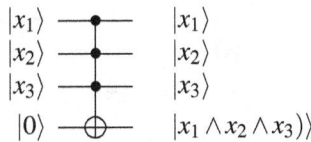

$$
\begin{array}{ll}
|x_1\rangle & |x_1\rangle \\
|x_2\rangle & |x_2\rangle \\
|x_3\rangle & |x_3\rangle \\
|0\rangle & |x_1 \wedge x_2 \wedge x_3\rangle)
\end{array}
$$

In the same way as for the Toffoli gate with two inputs, this function $f(x_1, x_2, x_3)$ is equal to 1 (i.e., $y = 1$) if and only if $x_1 = 1$, $x_2 = 1$, and $x_3 = 1$, otherwise $y = 0$.

Now, consider the case where we want to the design the same circuit but for a different unique input leading to an output 1. Consider the input $|101\rangle$ instead of $|111\rangle$. This can be obtained from the Toffoli gate by switching the inputs 0 before the Toffoli gate, i.e., the 2nd input. This can be done by adding a NOT operation using the X gate. Then, for this input, we must add another X gate after the Toffoli gate to ensure that the value of the output is the same as the one of the corresponding input. Then, we obtain the following circuit:

$$|x_1\rangle \;\text{———•———}\; |x_1\rangle$$
$$|x_2\rangle \;\text{—}\boxed{X}\text{—•—}\boxed{X}\text{—}\; |x_2\rangle$$
$$|x_3\rangle \;\text{———•———}\; |x_3\rangle$$
$$|0\rangle \;\text{———}\oplus\text{———}\; |x_1 \wedge x_2 \wedge x_3\rangle$$

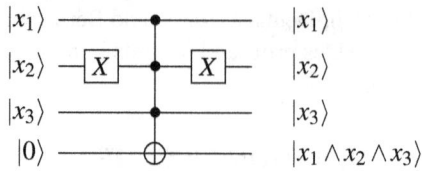

This circuit can be designed using the anti-control operation:

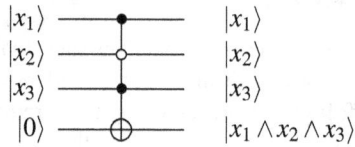

$$|x_1\rangle \;\text{——•——}\; |x_1\rangle$$
$$|x_2\rangle \;\text{——○——}\; |x_2\rangle$$
$$|x_3\rangle \;\text{——•——}\; |x_3\rangle$$
$$|0\rangle \;\text{——}\oplus\text{——}\; |x_1 \wedge x_2 \wedge x_3\rangle$$

The representation of this circuit can be simplified as follows using the symbol U_f:

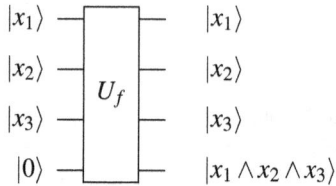

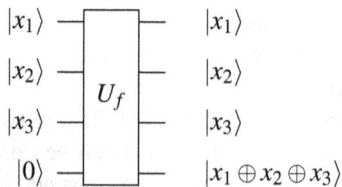

$$|x_1\rangle \;\text{—}\; |x_1\rangle$$
$$|x_2\rangle \quad U_f \quad |x_2\rangle$$
$$|x_3\rangle \;\text{—}\; |x_3\rangle$$
$$|0\rangle \;\text{—}\; |x_1 \wedge x_2 \wedge x_3\rangle$$

4.2.2 CIRCUIT 2: BALANCED CIRCUIT

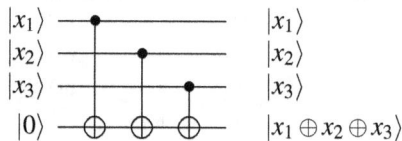

$$|x_1\rangle \;\text{—•————}\; |x_1\rangle$$
$$|x_2\rangle \;\text{———•——}\; |x_2\rangle$$
$$|x_3\rangle \;\text{—————•—}\; |x_3\rangle$$
$$|0\rangle \;\text{—}\oplus\text{—}\oplus\text{—}\oplus\text{—}\; |x_1 \oplus x_2 \oplus x_3\rangle$$

The representation of this circuit can be simplified as follows, using the symbol U_f:

$$|x_1\rangle \;\text{—}\; |x_1\rangle$$
$$|x_2\rangle \quad U_f \quad |x_2\rangle$$
$$|x_3\rangle \;\text{—}\; |x_3\rangle$$
$$|0\rangle \;\text{—}\; |x_1 \oplus x_2 \oplus x_3\rangle$$

4.2.3 PHASE ORACLE

The phase oracle is an oracle where the last input is set to $|-\rangle$. It has specific writing which helps to develop many quantum algorithms such as Deutsch's algorithm and Deutsch–Josza's algorithms.

An Oracle is given by the following circuit:

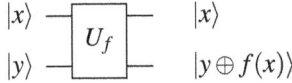

$$
\begin{array}{c}
|x\rangle \;\text{——}\; \boxed{U_f} \;\text{——}\; |x\rangle \\
|y\rangle \;\text{——}\; \phantom{\boxed{U_f}} \;\text{——}\; |y \oplus f(x)\rangle
\end{array}
$$

It can be written as:

$$U_f(|x\rangle|y\rangle) = |x\rangle|y \oplus f(x)\rangle$$

In case where $|y\rangle = |0\rangle$, this circuit represents the function of x since its output gives $|f(x)\rangle = |0 \oplus f(x)\rangle$, as shown by the following circuit.

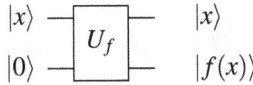

$$
\begin{array}{c}
|x\rangle \;\text{——}\; \boxed{U_f} \;\text{——}\; |x\rangle \\
|0\rangle \;\text{——}\; \phantom{\boxed{U_f}} \;\text{——}\; |f(x)\rangle
\end{array}
$$

This can be written as:

$$U_f(|x\rangle|0\rangle) = |x\rangle|f(x)\rangle \tag{4.1}$$

When $|y\rangle = |1\rangle$, we obtain $\overline{|f(x)\rangle}$ as an output, which can be represented by the following circuit:

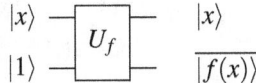

$$
\begin{array}{c}
|x\rangle \;\text{——}\; \boxed{U_f} \;\text{——}\; |x\rangle \\
|1\rangle \;\text{——}\; \phantom{\boxed{U_f}} \;\text{——}\; \overline{|f(x)\rangle}
\end{array}
$$

This circuit can be written as:

$$U_f(|x\rangle|1\rangle) = |x\rangle\overline{|f(x)\rangle} \tag{4.2}$$

For the Phase Oracle, consider the case where $|y\rangle = |-\rangle$. Then $U_f(|x-\rangle)$ is given as:

$$
\begin{aligned}
U_f(|x\rangle|-\rangle) &= U_f|x\rangle\frac{1}{\sqrt{2}}(|0\rangle - |1\rangle) \\
&= U_f\frac{1}{\sqrt{2}}(|x\rangle|0\rangle - |x\rangle|1\rangle) \\
&= \frac{1}{\sqrt{2}}(U_f(|x\rangle|0\rangle) - U_f(|x\rangle|1\rangle))
\end{aligned}
$$

From Equations (4.1) and (4.2), we can write:

$$U_f(|x\rangle|-\rangle) = \frac{1}{\sqrt{2}}(|x\rangle|f(x)\rangle - |x\rangle\overline{|f(x)\rangle}) \qquad (4.3)$$

Consider the cases where $|f(x)\rangle = |0\rangle$ and $|f(x)\rangle = |1\rangle$. In the case where $|f(x)\rangle = |0\rangle$, Equation (4.3) can be written as:

$$U_f(|x\rangle|-\rangle) = \frac{1}{\sqrt{2}}(|x\rangle|0\rangle - |x\rangle\overline{|0\rangle})$$

$$= \frac{1}{\sqrt{2}}(|x\rangle|0\rangle - |x\rangle|1\rangle)$$

$$= |x\rangle\frac{1}{\sqrt{2}}(|0\rangle - |1\rangle)$$

$$= |x\rangle|-\rangle$$

Therefore:

$$U_f(|x\rangle|-\rangle) = |x\rangle|-\rangle, \text{ if } |f(x)\rangle = |0\rangle \qquad (4.4)$$

Now, in the case where $|f(x)\rangle = |1\rangle$, Equation (4.3) can be written as:

$$U_f(|x\rangle|-\rangle) = \frac{1}{\sqrt{2}}(|x\rangle|1\rangle - |x\rangle\overline{|1\rangle})$$

$$= \frac{1}{\sqrt{2}}(|x\rangle|1\rangle - |x\rangle|0\rangle)$$

$$= |x\rangle\frac{1}{\sqrt{2}}(|1\rangle - |0\rangle)$$

$$= -|x\rangle\frac{1}{\sqrt{2}}(|0\rangle - |1\rangle)$$

$$= -|x\rangle|-\rangle$$

Therefore:

$$U_f(|x\rangle|-\rangle) = -|x\rangle|-\rangle, \text{ if } |f(x)\rangle = |1\rangle \qquad (4.5)$$

We can write Equations (4.4) and (4.5) in one Equation as follows:

$$U_f(|x\rangle|-\rangle) = (-1)^{f(x)}|x\rangle|-\rangle \qquad (4.6)$$

This can be represented by the following circuit called **Phase Oracle**:

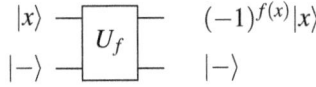

$$|x\rangle \quad \boxed{U_f} \quad (-1)^{f(x)}|x\rangle$$
$$|-\rangle \quad \qquad \quad |-\rangle$$

4.3 DEUTSCH'S ALGORITHM

4.3.1 CONSTANT AND BALANCED FUNCTIONS

The **Constant function** can be defined as the function for which all its outputs are the same and equal to 0 or to 1. For a one input function, we define two constant functions as follows:

x	$\rightarrow$	$f(x)$		x	$\rightarrow$	$f(x)$
0	$\rightarrow$	0	or	0	$\rightarrow$	1
1	$\rightarrow$	0		1	$\rightarrow$	1

The **Balanced function** can be defined as the function where the number of outputs that are equal to 0 is equal to the number of outputs that are equal to 1. For a one input function, we can define two Balanced functions as follows:

x	$\rightarrow$	$f(x)$		x	$\rightarrow$	$f(x)$
0	$\rightarrow$	0	or	0	$\rightarrow$	1
1	$\rightarrow$	1		1	$\rightarrow$	0

4.3.2 DEUTSCH'S CIRCUIT

The **Circuit of Deutsch** is given as follows [4]:

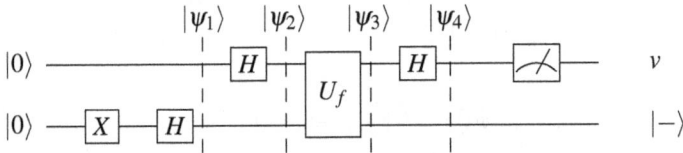

It can also be represented as follows:

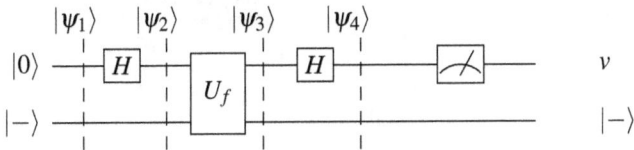

To find the output of Deutsch's algorithm, in the following, we calculate in each step the corresponding output $|\psi_1\rangle$, $|\psi_2\rangle$, $|\psi_3\rangle$, and $|\psi_4\rangle$. Thus,

$$|\psi_1\rangle = |0\rangle|-\rangle$$

$$\begin{aligned}|\psi_2\rangle &= H(|0\rangle)|-\rangle\\ &= \frac{1}{\sqrt{2}}(|0\rangle + |1\rangle)|-\rangle\\ &= \frac{1}{\sqrt{2}}(|0\rangle|-\rangle + |1\rangle|-\rangle)\end{aligned}$$

$$\begin{aligned}|\psi_3\rangle &= \frac{1}{\sqrt{2}}U_f(|0\rangle|-\rangle + |1\rangle|-\rangle)\\ &= \frac{1}{\sqrt{2}}(U_f(|0\rangle|-\rangle) + U_f(|1\rangle|-\rangle))\end{aligned}$$

From Equation (4.6), we can write:

$$\begin{aligned}|\psi_3\rangle &= \frac{1}{\sqrt{2}}((-1)^{f(0)}|0\rangle|-\rangle + (-1)^{f(1)}|1\rangle|-\rangle)\\ &= \frac{1}{\sqrt{2}}((-1)^{f(0)}|0\rangle + (-1)^{f(1)}|1\rangle)|-\rangle\end{aligned}$$

If $f(x)$ is constant and equal to 0, i.e., $f(0) = f(1) = 0$, thus,

$$\begin{aligned}|\psi_3\rangle &= \frac{1}{\sqrt{2}}((-1)^0|0\rangle|-\rangle + (-1)^0|1\rangle|-\rangle)\\ &= \frac{1}{\sqrt{2}}(|0\rangle + |1\rangle)|-\rangle\\ &= |+\rangle|-\rangle\end{aligned}$$

Then,

$$|\psi_4\rangle = H(|+\rangle)|-\rangle = |0\rangle|-\rangle$$

Therefore, the measurement $v = 0$.

If $f(x)$ is constant and equal to 1, i.e., $f(0) = f(1) = 1$, thus,

$$\begin{aligned}|\psi_3\rangle &= \frac{1}{\sqrt{2}}((-1)^1|0\rangle|-\rangle + (-1)^1|1\rangle|-\rangle)\\ &= -\frac{1}{\sqrt{2}}(|0\rangle + |1\rangle)|-\rangle\\ &= (-|+\rangle)|-\rangle\end{aligned}$$

Then,

$$|\psi_4\rangle = H(-|+\rangle)|-\rangle = -H(|+\rangle)|-\rangle = -|0\rangle|-\rangle$$

Therefore, the measurement $v = 0$.

If $f(x)$ is balanced and $f(0) = 0$ and $f(1) = 1$, thus,

$$|\psi_3\rangle = \frac{1}{\sqrt{2}}((-1)^0|0\rangle|-\rangle + (-1)^1|1\rangle|-\rangle)$$

$$= \frac{1}{\sqrt{2}}(|0\rangle - |1\rangle)|-\rangle$$

$$= |-\rangle|-\rangle$$

Then,

$$|\psi_4\rangle = H(|-\rangle)|-\rangle = |1\rangle|-\rangle$$

Therefore, the measurement $v = 1$.

If $f(x)$ is balanced and $f(0) = 1$ and $f(1) = 0$, therefore,

$$|\psi_3\rangle = \frac{1}{\sqrt{2}}((-1)^1|0\rangle|-\rangle + (-1)^0|1\rangle|-\rangle)$$

$$= \frac{1}{\sqrt{2}}(-|0\rangle + |1\rangle)|-\rangle$$

$$= -\frac{1}{\sqrt{2}}(|0\rangle - |1\rangle)|-\rangle$$

$$= (-|-\rangle)|-\rangle$$

Then,

$$|\psi_4\rangle = H(-|-\rangle)|-\rangle = -H(|-\rangle)|-\rangle = -|1\rangle|-\rangle$$

Therefore, the measurement $v = 1$:

$$v = \begin{cases} 0 & \text{if } f(x) = 0 \text{ or } f(x) = 1 \rightarrow f(x) \text{ is Constant} \\ 1 & \text{if } f(x) = x \text{ or } f(x) = \bar{x} \rightarrow f(x) \text{ is Balanced} \end{cases}$$

We conclude that if the function $f(x)$ is constant (i.e., $f(0) = f(1)$), then the output $v = 0$, and if the function $f(x)$ is balanced, then the output $v = 1$.

4.3.3 DEUTSCH'S QISKIT PROGRAM

The Qiskit program of the Deutsch algorithm is given as follows. In this program, we consider a balanced function $f(x) = \bar{x}$ which can be designed by a CNOT gate (line 8). The same program can be used to test constant functions, you just need to replace the function of line 8 of the following program by the identity gate `circuit.i(0)`, or just replace line 8 by a NOT gate (i.e., `circuit.x(0)`). These gates do not have any relation with the second output of the circuit. Thus, this output stays fixed to 0 (i.e., $f(x) = 0$).

```
1  import qiskit as q
2  simulator = q.Aer.get_backend('qasm_simulator')
```

```
3   circuit = q.QuantumCircuit(2,1)
4
5   circuit.h(0)
6   circuit.x(1)
7   circuit.h(1)
8   circuit.cx(0,1)
9   circuit.h(0)
10
11  circuit.measure(0, 0)
12
13  print(circuit.draw(output='text'))
14  job = q.execute(circuit, simulator, shots=1000)
15  result = job.result()
16  output = result.get_counts(circuit)
17  print(output)
18  import matplotlib.pyplot as plt
19  q.visualization.plot_histogram(output)
20  plt.show()
```

Note that the number of measurements is fixed to 1000 in the program (line 14). The result of this program is given by Figure 4.2. The bar shows a value of 1000, which is the number of times the value 1 is measured. This means that we obtained the value 1 with the probability of $1 = 1000/1000$. This implies that the function $f(x) = \bar{x}$ of line 8 is balanced.

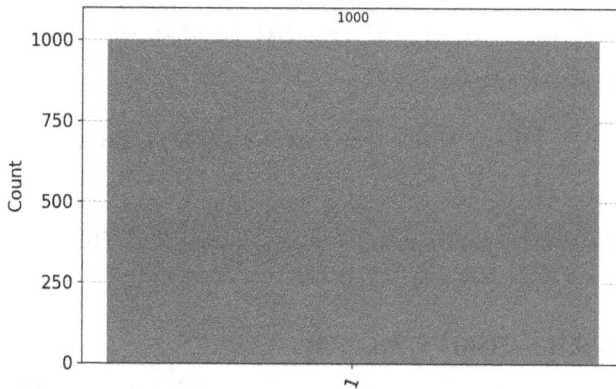

Figure 4.2 Histogram output of the Deutsch algorithm for a balanced function $|f(x)\rangle = \bar{x}$.

Figure 4.3 shows the obtained result for a function $f(x) = 0$, where line 8 is replaced by `circuit.x(0)`. The bar represents a value of 1000, which is the number of times the value 0 is measured. This means that we obtained the value 0 with the probability of $1 = 1000/1000$. This implies that the function $f(x) = 0$ is constant.

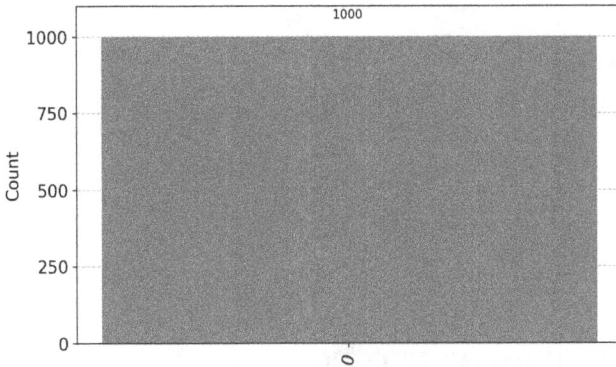

Figure 4.3 Histogram output of the Deutsch algorithm for $|f(x)\rangle = 0$.

4.4 DEUTSCH–JOZSA'S ALGORITHM

4.4.1 DEUTSCH–JOZSA'S CIRCUIT

The Deutsch–Jozsa's algorithm [4] is the generalisation of the Deutsch's algorithm. It considers many inputs instead of only one. The circuit of the Deutsch–Jozsa algorithm is given by the two following illustrations [4, 5]:

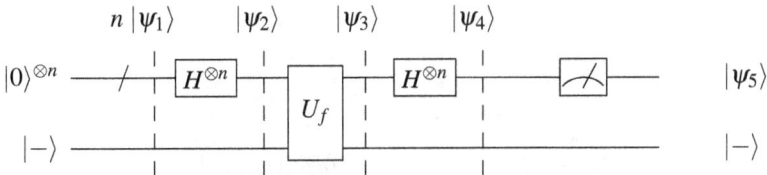

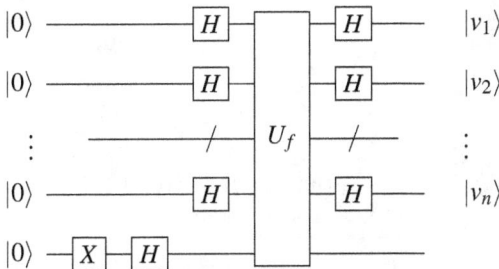

The following figure represents the Deutsch–Jozsa's circuit for the case of four inputs.

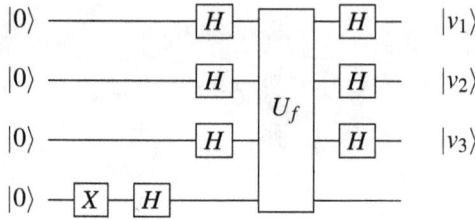

4.4.2 DEUTSCH–JOZSA'S EXECUTION

To find the output $|\psi_5\rangle$ of Deutsch's algorithm in the following, we calculate in each step the corresponding output $|\psi_1\rangle$, $|\psi_2\rangle$, $|\psi_3\rangle$, and $|\psi_4\rangle$. Then,

$$|\psi_1\rangle = |0\rangle^{\otimes n}|-\rangle$$

$$\begin{aligned}|\psi_2\rangle &= H^{\otimes n}(|0\rangle^{\otimes n})|-\rangle\\ &= H(|0\rangle)H(|0\rangle)...H(|0\rangle)|-\rangle\\ &= |++...+\rangle|-\rangle\end{aligned}$$

How to calculate $|++...+\rangle$?

$$|++...+\rangle = |+\rangle|+\rangle...|+\rangle$$
$$= \left(\frac{1}{\sqrt{2}}(|0\rangle + |1\rangle)\right)\left(\frac{1}{\sqrt{2}}(|0\rangle + |1\rangle)\right)...\left(\frac{1}{\sqrt{2}}(|0\rangle + |1\rangle)\right)$$

First, consider the case of two inputs, i.e., $n = 2$:

$$\begin{aligned}|++\rangle &= \left(\frac{1}{\sqrt{2}}(|0\rangle + |1\rangle)\right)\left(\frac{1}{\sqrt{2}}(|0\rangle + |1\rangle)\right)\\ &= \frac{1}{\sqrt{2^2}}((|0\rangle + |1\rangle)(|0\rangle + |1\rangle))\\ &= \frac{1}{\sqrt{2^2}}(|00\rangle + |01\rangle + |10\rangle + |11\rangle)\end{aligned}$$

Thus, we can write,

$$|++\rangle = \frac{1}{\sqrt{2^2}}\sum_{x\in\{0,1\}^2}|x\rangle$$

Then, consider the case of $n = 3$:

$$|{+}{+}{+}\rangle = \left(\frac{1}{\sqrt{2}}(|0\rangle + |1\rangle)\right)\left(\frac{1}{\sqrt{2}}(|0\rangle + |1\rangle)\right)\left(\frac{1}{\sqrt{2}}(|0\rangle + |1\rangle)\right)$$

$$= \frac{1}{\sqrt{2^2}}((|0\rangle + |1\rangle)(|0\rangle + |1\rangle)(|0\rangle + |1\rangle))$$

$$= \frac{1}{\sqrt{2^2}}(|000\rangle + |001\rangle + |010\rangle + |011\rangle + |100\rangle + |101\rangle + |110\rangle + |111\rangle)$$

Thus, we can write,

$$|{+}{+}{+}\rangle = \frac{1}{\sqrt{2^3}} \sum_{x \in \{0,1\}^3} |x\rangle$$

We conclude that for any value of n,

$$\boxed{H^{\otimes n}(|0\rangle^{\otimes n}) = \frac{1}{\sqrt{2^n}} \sum_{x \in \{0,1\}^n} |x\rangle} \tag{4.7}$$

Therefore,

$$|\psi_2\rangle = \frac{1}{\sqrt{2^n}} \sum_{x \in \{0,1\}^n} |x\rangle |{-}\rangle$$

Then, we calculate $|\psi_3\rangle = U_f(|\psi_2\rangle)$, as follows:

$$|\psi_3\rangle = U_f(|\psi_2\rangle)$$

$$= U_f \left(\frac{1}{\sqrt{2^n}} \sum_{x \in \{0,1\}^n} |x\rangle |{-}\rangle \right)$$

$$= \frac{1}{\sqrt{2^n}} U_f \sum_{x \in \{0,1\}^n} |x\rangle |{-}\rangle$$

$$= \frac{1}{\sqrt{2^n}} \sum_{x \in \{0,1\}^n} U_f |x\rangle |{-}\rangle$$

$$= \frac{1}{\sqrt{2^n}} \sum_{x \in \{0,1\}^n} (-1)^{f(x)} |x\rangle |{-}\rangle \quad \text{from Equation (4.6)}$$

Consider:

$$|\psi_{3'}\rangle = \frac{1}{\sqrt{2^n}} \sum_{x \in \{0,1\}^n} (-1)^{f(x)} |x\rangle$$

Then,

$$|\psi_3\rangle = |\psi_{3'}\rangle |{-}\rangle$$

Consider $|\psi_4\rangle = |\psi_{4'}\rangle|-\rangle$. Then,

$$|\psi_4\rangle = |\psi_{4'}\rangle|-\rangle$$
$$= H^{\otimes n}(|\psi_{3'}\rangle)|-\rangle$$
$$= H^{\otimes n}\left(\frac{1}{\sqrt{2^n}}\sum_{x\in\{0,1\}^n}(-1)^{f(x)}|x\rangle\right)|-\rangle$$
$$= \left(\frac{1}{\sqrt{2^n}}\sum_{x\in\{0,1\}^n}(-1)^{f(x)}H^{\otimes n}|x\rangle\right)|-\rangle$$
$$= \left(\frac{1}{\sqrt{2^n}}\sum_{x\in\{0,1\}^n}(-1)^{f(x)}(H|x_1\rangle H|x_2\rangle...H|x_{n-1}\rangle)\right)|-\rangle$$

To calculate the value of $H(|x_i\rangle)$, let us recall the values of $H(|0\rangle)$ and $H(|1\rangle)$:

$$H(|0\rangle) = \frac{1}{\sqrt{2}}(|0\rangle + |1\rangle)$$

$$H(|1\rangle) = \frac{1}{\sqrt{2}}(|0\rangle - |1\rangle)$$

These two equations can be written with only one equation as follows:

$$\boxed{H(|x_i\rangle) = \frac{1}{\sqrt{2}}(|0\rangle + (-1)^{x_i}|1\rangle)} \tag{4.8}$$

To calculate $H^{\otimes n}|x\rangle$ in ψ_4, let us consider the case where $|x\rangle = |x_0 x_1 x_2\rangle$. Then,

$$H^{\otimes 3}|x\rangle = H^{\otimes 3}|x_0 x_1 x_2\rangle$$
$$= H(|x_0\rangle)H(|x_1\rangle)H(|x_2\rangle)$$
$$= \frac{1}{\sqrt{2}}(|0\rangle + (-1)^{x_0}|1\rangle)\frac{1}{\sqrt{2}}(|0\rangle + (-1)^{x_1}|1\rangle)\frac{1}{\sqrt{2}}(|0\rangle + (-1)^{x_2}|1\rangle)$$
$$= \frac{1}{\sqrt{2^3}}(|0\rangle + (-1)^{x_0}|1\rangle)(|0\rangle + (-1)^{x_1}|1\rangle)(|0\rangle + (-1)^{x_2}|1\rangle)$$
$$= \frac{1}{\sqrt{2^3}}(|000\rangle + (-1)^{x_0}|100\rangle + (-1)^{x_1}|010\rangle + (-1)^{x_2}|001\rangle +$$
$$(-1)^{x_0}(-1)^{x_1}|110\rangle + (-1)^{x_0}(-1)^{x_2}|101\rangle + (-1)^{x_1}(-1)^{x_2}|011\rangle +$$
$$(-1)^{x_0}(-1)^{x_1}(-1)^{x_2}|111\rangle)$$
$$= \frac{1}{\sqrt{2^3}}(|000\rangle + (-1)^{x_0}|100\rangle + (-1)^{x_1}|010\rangle + (-1)^{x_2}|001\rangle +$$
$$(-1)^{x_0+x_1}|110\rangle + (-1)^{x_0+x_2}|101\rangle + (-1)^{x_1+x_2}|011\rangle +$$
$$(-1)^{x_0+x_1+x_2}|111\rangle)$$
$$= \frac{1}{\sqrt{2^3}}((-1)^{x\cdot000}|000\rangle + (-1)^{x\cdot100}|100\rangle + (-1)^{x\cdot010}|010\rangle + (-1)^{x\cdot001}|001\rangle +$$
$$(-1)^{x\cdot011}|011\rangle + (-1)^{x\cdot101}|101\rangle + (-1)^{x\cdot110}|110\rangle +$$
$$(-1)^{x\cdot111}|111\rangle$$

Therefore,

$$H^{\otimes 3}|x\rangle = \frac{1}{\sqrt{2^3}} \left(\sum_{z \in \{0,1\}^3} (-1)^{xz}|z\rangle \right)$$

This can be generalised to n as follows:

$$H^{\otimes n}|x\rangle = \frac{1}{\sqrt{2^n}} \left(\sum_{z \in \{0,1\}^n} (-1)^{xz}|z\rangle \right) \qquad (4.9)$$

In case where $|x\rangle = |00...0\rangle$, we can retrieve the formula of Equation (4.7):

$$H^{\otimes n}(|00...0\rangle) = H^{\otimes n}(|0\rangle^{\otimes n}) = \frac{1}{\sqrt{2^n}} \sum_{x \in \{0,1\}^n} |x\rangle$$

Then,

$$
\begin{aligned}
|\psi_{4'}\rangle &= \frac{1}{\sqrt{2^n}} \sum_{x \in \{0,1\}^n} (-1)^{f(x)} H^{\otimes n}|x\rangle \\
&= \frac{1}{\sqrt{2^n}} \sum_{x \in \{0,1\}^n} (-1)^{f(x)} \frac{1}{\sqrt{2^n}} \sum_{z \in \{0,1\}^n} (-1)^{xz}|z\rangle \\
&= \sum_{x \in \{0,1\}^n} \frac{1}{2^n} \sum_{z \in \{0,1\}^n} (-1)^{f(x)}(-1)^{xz}|z\rangle
\end{aligned}
$$

Therefore,

$$|\psi_{4'}\rangle = \sum_{z \in \{0,1\}^n} \left(\frac{1}{2^n} \sum_{x \in \{0,1\}^n} (-1)^{f(x)+xz} \right) |z\rangle$$

Consider the output state $|00...0\rangle$, thus,

$$
\begin{aligned}
|\psi_{4'}\rangle &= \left(\frac{1}{2^n} \sum_{x \in \{0,1\}^n} (-1)^{f(x)+x(00...0)} \right) |00...0\rangle \\
&= \left(\frac{1}{2^n} \sum_{x \in \{0,1\}^n} (-1)^{f(x)} \right) |00...0\rangle \\
&= \alpha|00...0\rangle
\end{aligned}
$$

α is the coefficient allowing to calculate the probability $|\alpha|^2$ to measure the state $|00...0\rangle$. The value of this coefficient depends on the function $f(x)$. Let us consider that $f(x)$ is constant and equal to 0, i.e., $f(x) = 0, \forall x$, then,

$$\alpha = \frac{1}{2^n} \sum_{x \in \{0,1\}^n} (-1)^{f(x)}$$

$$= \frac{1}{2^n} \sum_{x \in \{0,1\}^n} (-1)^0$$

$$= \frac{1}{2^n} \sum_{x \in \{0,1\}^n} 1$$

$$= \frac{2^n}{2^n}$$

$$= 1$$

Therefore, if $f(x) = 0$, then the output $|\psi_{4'}\rangle = |00...0\rangle$.

If we consider that the function $f(x)$ is constant and equal to 1, i.e., $f(x) = 1, \forall x$, then,

$$\alpha = \frac{1}{2^n} \sum_{x \in \{0,1\}^n} (-1)^{f(x)}$$

$$= \frac{1}{2^n} \sum_{x \in \{0,1\}^n} (-1)^1$$

$$= \frac{1}{2^n} \sum_{x \in \{0,1\}^n} -1$$

$$= -\frac{2^n}{2^n}$$

$$= -1$$

Therefore, if $f(x) = 1$ then the output $|\psi_{4'}\rangle = |00...0\rangle$.

We conclude from these two last equations where $f(x) = 0$ or $f(x) = 1$ that if the function $f(x)$ is constant, then the output of the algorithm of Deutsch–Jozsa is $|00...0\rangle$, or in another way if the output of the Deutsch–Jozsa's algorithm is equal to $|00...0\rangle$, then the function $f(x)$ is **constant**.

Now, consider another case where the output has $n_1 = 2^n/2$ of 1s and $n_2 = 2^n/2$ of 0s, then,

$$|\psi_{4'}\rangle = \left(\frac{1}{2^n} \sum_{x \in \{0,1\}^{n_1}} (-1)^{f(x)=0} + \sum_{x \in \{0,1\}^{n_2}} (-1)^{f(x)=1} \right) |00...0\rangle$$

$$= \left(\frac{1}{2^n} \sum_{x \in \{0,1\}^{n_1}} (-1)^0 + \sum_{x \in \{0,1\}^{n_2}} (-1)^1 \right) |00...0\rangle$$

$$= \left(\frac{1}{2^n} \sum_{x \in \{0,1\}^{n_1}} 1 + \sum_{x \in \{0,1\}^{n_2}} -1 \right) |00...0\rangle$$

$$= \left(\frac{1}{2^n} \left(\frac{2^n}{2} - \frac{2^n}{2} \right) \right) |00...0\rangle$$

$$= 0|00...0\rangle$$

Note that $\{0,1\}^{n_1} \cup \{0,1\}^{n_2} = \{0,1\}^n$ and $\{0,1\}^{n_1} \cap \{0,1\}^{n_2} = \emptyset$

We conclude from this equation that if the function $f(x)$ is balanced then the output of the algorithm of Deutsch–Jozsa can never be equal to $|00...0\rangle$, or in another way, if the output of the Deutsch–Jozsa's algorithm is different from $|00...0\rangle$ then the function $f(x)$ is **balanced**.

4.5 BERNSTEIN–VAZIRANI ALGORITHM

4.5.1 PROBLEM DEFINITION

The Bernstein–Vazirani algorithm [6] is a quantum algorithm that efficiently solves a classical problem known as the hidden oracle problem. It was developed by Ethan Bernstein and Umesh Vazirani in 1993. This algorithm is notable for its ability to determine a hidden binary code in a single iteration, thus providing significant acceleration compared to classical methods, which determine the hidden code in n iterations. It calculates the AND operation with each bit of the hidden code and 1. If the AND result is equal to 0, then the corresponding value in the code is equal to 0 too, otherwise, the corresponding value in the code is equal to 1.

This problem can be defined mathematically as follows. Consider a hidden code $s = (s_0 s_1 ... s_{n-1})$ and a function $f : \{0,1\}^n \to \{0,1\}$ such that $f(x) = x.s \ (mod \ 2)$:

$$f(100...0) = (100...0) \cdot s = s_0$$
$$f(010...0) = (010...0) \cdot s = s_1$$

$$...$$

$$f(000...1) = (100...0) \cdot s = s_{n-1}$$

For example, in case where the code $s = (10110)$, we obtain:

$$f(10000) = (10000) \cdot (10110) = 1$$
$$f(01000) = (01000) \cdot (10110) = 0$$
$$f(00100) = (00100) \cdot (10110) = 1$$
$$f(00010) = (00010) \cdot (10110) = 1$$
$$f(00001) = (00001) \cdot (10110) = 0$$

4.5.2 BERNSTEIN–VAZIRANI'S CIRCUIT

The quantum circuit allows resolving this problem is exactly the same as the one of Deutsch–Jozsa presented in the previous sections. It is given as follows.

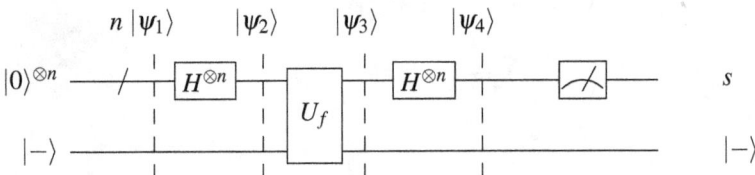

4.5.3   BERNSTEIN–VAZIRANI'S EXECUTION

Calculate its different states $|\psi_i\rangle$ of this circuit:

$$|\psi_1\rangle = |0\rangle^{\otimes n}|-\rangle$$

$$|\psi_2\rangle = H^{\otimes n}|0\rangle^{\otimes n}|-\rangle$$
$$= \frac{1}{\sqrt{2^n}} \sum_{x\in\{0,1\}^n} |x\rangle|-\rangle$$

$$|\psi_3\rangle = U_f|\psi_2\rangle$$
$$= U_f \left(\frac{1}{\sqrt{2^n}} \sum_{x\in\{0,1\}^n} |x\rangle|-\rangle \right)$$
$$= \frac{1}{\sqrt{2^n}} \sum_{x\in\{0,1\}^n} U_f|x\rangle|-\rangle$$
$$= \frac{1}{\sqrt{2^n}} \sum_{x\in\{0,1\}^n} (-1)^{f(x)}|x\rangle|-\rangle$$
$$= \frac{1}{\sqrt{2^n}} \sum_{x\in\{0,1\}^n} (-1)^{x\cdot s}|x\rangle|-\rangle$$

$$|\psi_4\rangle = H^{\otimes n} \left(\frac{1}{\sqrt{2^n}} \sum_{x\in\{0,1\}^n} (-1)^{x\cdot s}|x\rangle \right) |-\rangle$$
$$= \left(\frac{1}{\sqrt{2^n}} \sum_{x\in\{0,1\}^n} (-1)^{x\cdot s} H^{\otimes n}|x\rangle \right) |-\rangle$$
$$= \left(\frac{1}{\sqrt{2^n}} \sum_{x\in\{0,1\}^n} (-1)^{x\cdot s} \frac{1}{\sqrt{2^n}} \sum_{z\in\{0,1\}^n} (-1)^{x\cdot z}|z\rangle \right) |-\rangle$$
$$= \left(\frac{1}{\sqrt{2^n}} \frac{1}{\sqrt{2^n}} \sum_{x\in\{0,1\}^n} \sum_{z\in\{0,1\}^n} (-1)^{x\cdot s}(-1)^{x\cdot z}|z\rangle \right) |-\rangle$$
$$= \left(\frac{1}{\sqrt{2^n}} \frac{1}{\sqrt{2^n}} \sum_{x\in\{0,1\}^n} \sum_{z\in\{0,1\}^n} (-1)^{x\cdot s + x\cdot z}|z\rangle \right) |-\rangle$$
$$= \left(\frac{1}{2^n} \sum_x \sum_z (-1)^{x\cdot(s+z)}|z\rangle \right) |-\rangle$$

Calculate the coefficient to obtain the state s, i.e., $z = s$:

$$|\psi_4\rangle = \left(\frac{1}{2^n}\sum_x (-1)^{x\cdot(s+s)}|s\rangle\right)|-\rangle$$

$$= \left(\frac{1}{2^n}\sum_x (-1)^{x\cdot(00...0)}|s\rangle\right)|-\rangle$$

$$= \left(\frac{1}{2^n}\sum_x (-1)^0|s\rangle\right)|-\rangle$$

$$= \left(\frac{1}{2^n}\sum_x 1|s\rangle\right)|-\rangle$$

$$= \left(\frac{2^n}{2^n}|s\rangle\right)|-\rangle$$

$$= |s\rangle|-\rangle$$

Note that $s+s$ represents $s \oplus s$, i.e., $s\ XOR\ s$, and it is known that $a \oplus b = 0^n = (0...0)$ if and only if $a = b$. This means that the probability to obtain the state $|s\rangle$ is equal to $1^2 = 1$. Thus, the probability of obtaining the other states is equal to 0. Therefore, the proposed circuit allows finding the code s.

The difference with Deutsch–Jozsa is based on the Oracle (U_f). For instance, the Oracle corresponding to the code $s = (10110)$ can be designed as follows:

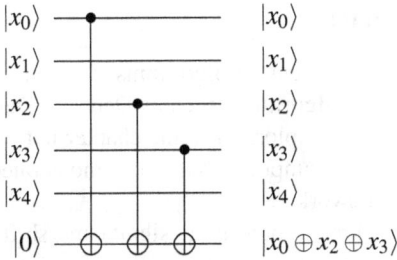

4.5.4 BERNSTEIN–VAZIRANI'S EXAMPLE

Consider as an example the following circuit, where $s = (101)$, and let calculate its different states $|\psi_i\rangle$:

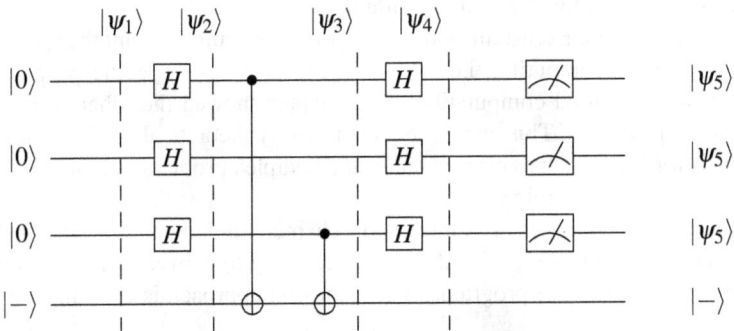

$$|\psi_1\rangle = |000\rangle|-\rangle$$

$$
\begin{aligned}
|\psi_2\rangle &= H.H.H|000\rangle|-\rangle \\
&= |+++\rangle|-\rangle \\
&= (|000\rangle + |001\rangle + |010\rangle + |011\rangle + |100\rangle + |101\rangle + |110\rangle + |111\rangle)|-\rangle
\end{aligned}
$$

$$
\begin{aligned}
|\psi_3\rangle &= CX_{2,3}CX_{0,3}|\psi_2\rangle \\
&= CX_{2,3}CX_{0,3}|+++\rangle|-\rangle \\
&= |-+-\rangle|-\rangle \\
&= (|000\rangle - |001\rangle + |010\rangle - |011\rangle - |100\rangle + |101\rangle - |110\rangle + |111\rangle)|-\rangle
\end{aligned}
$$

$$
\begin{aligned}
|\psi_4\rangle &= H.H.H|-+-\rangle|-\rangle \\
&= |101\rangle|-\rangle
\end{aligned}
$$

After the measurement, we obtain the output $s = 101$.

4.6 CONCLUSION

This chapter examined quantum algorithms, focusing primarily on the significant capabilities of seminal algorithms such as Deutsch, Deutsch–Jozsa, and Bernstein–Vazirani. Through these explorations, the chapter not only discussed the theoretical foundations of quantum computing but also demonstrated their practical applications using the Qiskit framework.

Quantum algorithms represent a significant shift from classical computing paradigms. By utilising the unique properties of quantum mechanics, such as superposition, entanglement, and quantum interference, these algorithms achieve computational efficiencies that are unattainable by their classical counterparts. The Deutsch and Deutsch–Jozsa algorithms, for example, demonstrate the potential of quantum computing to execute tasks with extraordinary speed, solving problems in a fraction of the time required by classical algorithms.

Furthermore, the Bernstein–Vazirani algorithm exemplifies another powerful aspect of quantum computing, i.e., its ability to determine specific properties of a function with minimal computation. This chapter showed the inherent parallelism of quantum processes. This capability is not merely theoretical but has practical implications for fields such as cryptography and complex problem-solving, from cybersecurity to molecular biology.

The practical exercises presented in this chapter, implemented via the Qiskit platform, aim to equip readers with the skills necessary to convert quantum computing concepts into executable programs. This hands-on approach is crucial, as it bridges the gap between theoretical knowledge and practical application, empowering enthusiasts and professionals alike to explore and innovate within the field of quantum computing.

As we look to the future, the integration of quantum computing into mainstream technology continues to present both challenges and opportunities. The scalability of quantum systems, error rates, and coherence times are among the technical hurdles that need further addressing. However, ongoing developments in quantum hardware and software, exemplified by platforms like Qiskit, promise a future where these challenges are increasingly surmountable.

In conclusion, the investigation of quantum algorithms opens up new horizons in computing, offering unprecedented opportunities for solving some of the most complex problems facing the world today. The knowledge and skills imparted in this chapter are foundational, paving the way for further research and development in this exciting and rapidly evolving field. As quantum technologies continue to mature, they will undoubtedly play a crucial role in shaping the future of computation, potentially transforming a range of industries and disciplines.

REFERENCES

1. Sanders, J.W., Zuliani, P.: Quantum programming. In: Backhouse, R., Oliveira, J.N. (eds.), Mathematics of Program Construction, pp. 80–99. Springer, Berlin, Heidelberg (2000).

2. Yavuz, D.D., Yadav, A.: Simulation of quantum algorithms using classical probabilistic bits and circuits (2023). https://arxiv.org/abs/2307.14452

3. Javadi-Abhari, A., Treinish, M., Krsulich, K., Wood, C.J., Lishman, J., Gacon, J., Martiel, S., Nation, P.D., Bishop, L.S., Cross, A.W., Johnson, B.R., Gambetta, J.M.: Quantum computing with Qiskit (2024). https://arxiv.org/abs/2405.08810

4. David, D., Richard, J.: Rapid solutions of problems by quantum computation. In: Proceedings of the Royal Society of London A. **439** (1907), 553–558 (1992).

5. Collins, D., Kim, K.W., Holton, W.C.: Deutsch-jozsa algorithm as a test of quantum computation. Physical Review A **58**(3), 1633–1636 (1998). https://doi.org/10.1103/physreva.58.r1633

6. Bernstein, E., Vazirani, U.: Quantum complexity theory. SIAM Journal on Computing **26** (5), 1411–1473 (1997). https://doi.org/10.1137/S0097539796300921.

5 Grover's Algorithm
Quantum Brute-Force Search

Ahcene Bounceur, Mohammad Hammoudeh,
Bamidele Adebisi, and Abdelkader Laouid

5.1 INTRODUCTION

This chapter presents an in-depth exploration of Grover's algorithm [1–3], a quantum algorithm known for its ability to search unsorted databases with remarkable efficiency, e.g., brute-force search, a task that can be extremely time-consuming for classical computers. Grover's algorithm offers a quantum-powered solution that significantly reduces the computational resources required to perform such searches, showcasing its ability in optimisation and efficiency. Through a mixture of theoretical elucidation and practical illustration, readers will gain a comprehensive understanding of this revolutionary algorithm.

The chapter starts with a thorough presentation of Grover's algorithm, examining its theoretical framework and explaining the fundamental principles that underlie its functionality. To facilitate understanding, a geometric representation is provided, offering readers a visual aid to intuitively understand the operations of the algorithm.

To facilitate a deeper understanding of Grover's algorithm, an illustrative example is presented, which serves as a recurring reference point throughout the chapter. This example is systematically utilised to present each step of the algorithm, ensuring clarity and coherence in the presented explanations.

Transitioning to the quantum domain, the chapter explores the construction of the quantum circuit responsible for implementing Grover's algorithm. Each component of this circuit is explained in details with a step-by-step breakdown offered to highlight the relevant quantum mechanical principles.

The chapter concludes with the practical application of Grover's algorithm using the Qiskit programming framework. Beginning with the implementation of the algorithm for the case of $n = 2$, readers are provided with a practical demonstration of its functionality. Furthermore, the chapter extends its exploration to showcase the algorithm's scalability for any given value of n, underlying its versatility and real-world applicability.

DOI: 10.1201/9781003475286-5

5.2   BRUTE-FORCE SEARCH PROBLEM

The main objective of Grover's algorithm is to find a hidden binary code, such as in the **Brute-force search** illustrated in Figure 5.1. This problem can be defined as follows. Given a function $f : \{0,1\}^n \to \{0,1\}$, find x, s.t. $f(x) = 1$, knowing that there is only one $x = x^*$ for which $f(x) = 1$.

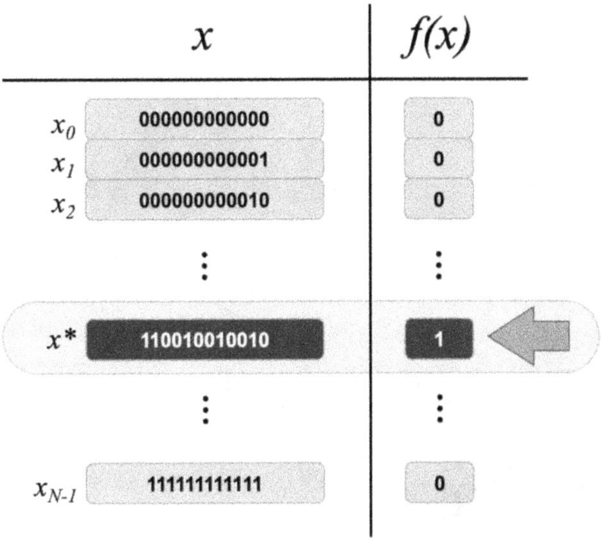

x		$f(x)$
x_0	000000000000	0
x_1	000000000001	0
x_2	000000000010	0
	$\vdots$	$\vdots$
x^*	110010010010	1
	$\vdots$	$\vdots$
x_{N-1}	111111111111	0

Figure 5.1   Finding a password: searching for a specific binary sequence.

This problem is known as "looking for a needle in a haystack" since its complexity with a classical computer is $O(2^n)$, i.e., exponential. Section 5.4.4 shows that its complexity with a quantum computer is reduced to $O(\sqrt{n})$.

5.3   BASIC CONCEPTS

Given $A = \{x \in \{0,1\}^n : f(x) = 0\}$ and $B = \{x \in \{0,1\}^n : f(x) = 1\}$, for $a = |A|$ and $b = |B|$. Consider $|A\rangle$ a superposition in A, $|B\rangle$ a superposition in B and $|\psi_0\rangle$ a superposition in A and B $(A \cup B)$. Then, we can write:

$$|A\rangle = \frac{1}{\sqrt{a}} \sum_{x \in A} |x\rangle$$

$$|B\rangle = \frac{1}{\sqrt{b}} \sum_{x \in B} |x\rangle$$

$$|h\rangle = \frac{1}{\sqrt{N}} \sum_{x \in \{0,1\}^n} |x\rangle$$

where, $N = 2^n$.

The superposition $|h\rangle$ can be written with respect to $|A\rangle$ and $|B\rangle$ as follows:

$$|h\rangle = \frac{1}{\sqrt{N}} \sum_{x \in A} |x\rangle + \frac{1}{\sqrt{N}} \sum_{x \in B} |x\rangle$$

$$= \frac{\sqrt{a}}{\sqrt{N}} \left(\frac{1}{\sqrt{a}} \sum_{x \in A} |x\rangle \right) + \frac{\sqrt{b}}{\sqrt{N}} \left(\frac{1}{\sqrt{b}} \sum_{x \in B} |x\rangle \right) \qquad (5.1)$$

$$= \frac{\sqrt{a}}{\sqrt{N}} |A\rangle + \frac{\sqrt{b}}{\sqrt{N}} |B\rangle$$

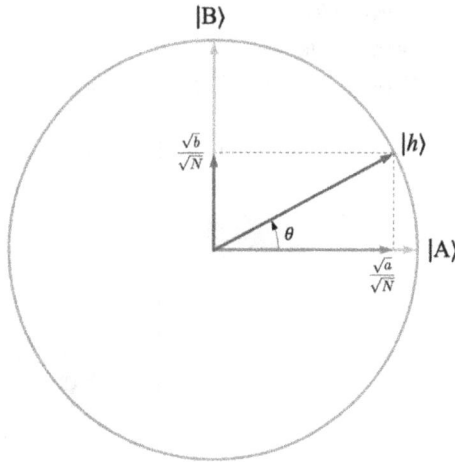

Figure 5.2 $|h\rangle$ as a function of $|A\rangle$ and $|B\rangle$.

Figure 5.2 illustrates this situation. We can write $|h\rangle$ as a function of θ as follows:

$$|h\rangle = \frac{\sqrt{a}}{\sqrt{N}} |A\rangle + \frac{\sqrt{b}}{\sqrt{N}} |B\rangle$$

$$= \cos\theta |A\rangle + \sin\theta |B\rangle \qquad (5.2)$$

$$\implies \theta = \frac{\sqrt{b}}{\sqrt{N}} \qquad (5.3)$$

based on the small angle approximation ($\sin\theta \approx \theta$).

5.4 GROVER'S ALGORITHM

Given a function $f : \{0,1\}^n \to \{0,1\}$, we need to find x, s.t. $f(x) = 1$, knowing that there is only one $x = x^*$ such that $f(x^*) = 1$.

This example serves to clarify the different theoretical concepts that are presented in this chapter. For this purpose, consider an example with three qubits, where $x^* = |101\rangle$ is the input such that $f(x^*) = 1$, and for all the other qubits $x \neq x^*$, $f(x) = 0$. Thus,

$$A = \{101\}$$

and,

$$B = \{000, 001, 010, 011, 100, 110, 111\}$$

Then, $a = |A| = 1$ and $b = |B| = 7$. Note that, $|A\rangle = |x^*\rangle = |101\rangle$.

The steps of the Grover's algorithm are given as follows:

- Step 1: Superposition (Section 5.4.1)
- Step 2: Phase inversion (Section 5.4.2)
- Step 3: Inversion about the mean (Section 5.4.3)
- Repeat k times from Step 2

where k is given in Section 5.4.4. These steps are detailed in the following sections.

5.4.1 THE SUPERPOSITION

Considered an initialisation step, the superposition $|\psi_0\rangle$ can be obtained by applying the Hadamard gate to the state $|0\rangle^{\otimes n}$ [4,5]. Then, we can write $|\psi_0\rangle$ as follows:

$$\begin{aligned}
|\psi_0\rangle = |h\rangle &= H^{\otimes n}|0\rangle^{\otimes n} \\
&= \frac{1}{\sqrt{N}} \sum_{x \in \{0,1\}^n} |x\rangle \\
&= \frac{1}{\sqrt{8}} (|000\rangle + |001\rangle + |010\rangle + |011\rangle + |100\rangle + |101\rangle + |110\rangle + |111\rangle) \\
&= \frac{1}{2\sqrt{2}} (|000\rangle + |001\rangle + |010\rangle + |011\rangle + |100\rangle + |101\rangle + |110\rangle + |111\rangle)
\end{aligned} \tag{5.4}$$

The superposition state $|\psi_0\rangle$ is shown by Figure 5.3. As we can see,

$$\alpha_{i, i=0,\dots,7} = \frac{1}{2\sqrt{2}}$$

The objective is to highlight the red state $|101\rangle$. To do this, we need to find a way to increase the value of the factor $\alpha^* = \alpha_5$, i.e., increase the probability $|\alpha_5|^2$ to obtain the state $|101\rangle$ and to reduce the value of the other factors $\alpha_{i, i \neq 5}$, i.e., reduce the probability $|\alpha_{i, i \neq 5}|^2$ to obtain the other states. In other terms, we need to find the quantum circuit to transform the factor α_5 from the value $\frac{1}{2\sqrt{2}}$ to ≈ 1 and the factors $\alpha_{i, i \neq 5}$ from the value $\frac{1}{2\sqrt{2}}$ to ≈ 0.

The superposition $|\psi_0\rangle$ as a function of $|A\rangle$ and $|B\rangle$ is given by:

$$\begin{aligned}
|\psi_0\rangle &= \alpha|A\rangle + \beta|B\rangle \\
&= \alpha|A\rangle + \alpha^*|x^*\rangle \\
&= \sqrt{\frac{1}{8}}|A\rangle + \sqrt{\frac{7}{8}}|x^*\rangle \\
&= \frac{1}{2\sqrt{2}}|A\rangle + \frac{\sqrt{7}}{2\sqrt{2}}|x^*\rangle
\end{aligned} \tag{5.5}$$

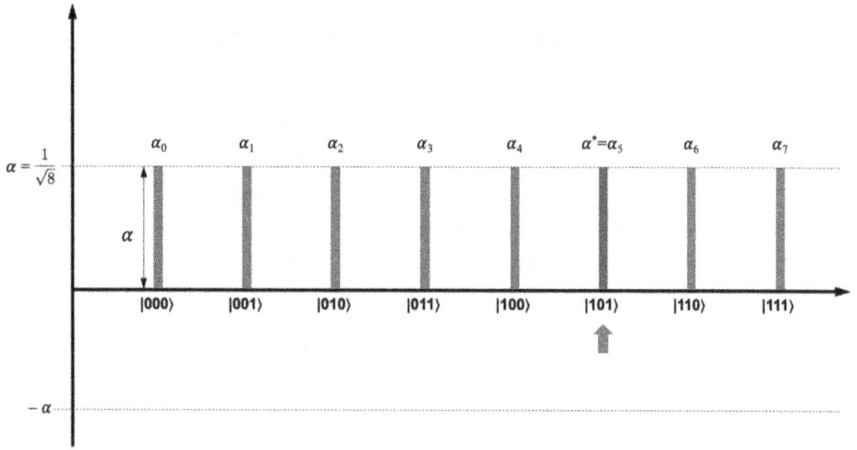

Figure 5.3   Superposition of three bits: State $|\psi_0\rangle$.

Figure 5.4 illustrates the representation of $|\psi_0\rangle$ as a function of $|A\rangle$ and $|B\rangle = |x^*\rangle$.

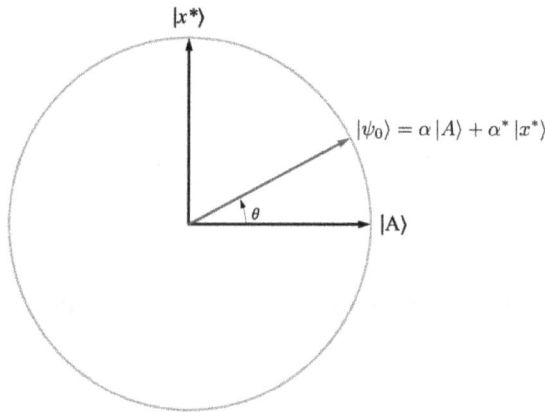

Figure 5.4 $|\psi_0\rangle$ as a function of $|A\rangle$ and $|x^*\rangle$.

5.4.2   PHASE INVERSION

Also known as "Oracle construction step", the phase inversion can be obtained from a circuit called an oracle, which has $n+1$ inputs. The first n inputs are set to the superposition by adding a Hadamard gate to each input. The input states can be represented by $|+\rangle$. The last $(n+1)$th input is set to the state $|-\rangle$ which is obtained from an input $|0\rangle$ by adding two gates, the X gate followed by the H gate. In this

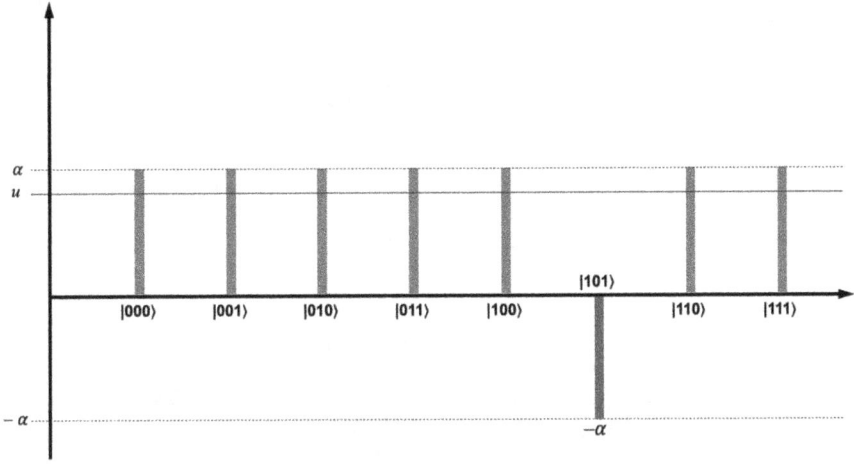

Figure 5.5   The coefficients of the state $|\psi_1\rangle$ after the phase inversion step.

case, we can write the output $|\psi_1\rangle$ of the gate as follows:

$$
\begin{aligned}
|\psi_1\rangle = U_f(|\psi_0\rangle|-\rangle) &= \frac{1}{\sqrt{N}} \sum_{x\in\{0,1\}^n} U_f(|x\rangle|-\rangle) \\
&= \frac{1}{\sqrt{N}} \sum_{x\in\{0,1\}^n} (-1)^{f(x)}|x\rangle|-\rangle \\
&= \frac{1}{\sqrt{N}} \left(\sum_{x\in A} (-1)^{f(x)}|x\rangle + \sum_{x\in B} (-1)^{f(x)}|x\rangle \right)|-\rangle \quad (5.6) \\
&= \frac{1}{\sqrt{N}} \left(\sum_{x\in A} |x\rangle - \sum_{x\in B} |x\rangle \right)|-\rangle \\
&= \frac{1}{\sqrt{N}} \left(\sum_{x\neq x^*} |x\rangle - |x^*\rangle \right)|-\rangle
\end{aligned}
$$

$$
|\psi_1\rangle = \frac{1}{2\sqrt{2}} \left(|000\rangle + |001\rangle + |010\rangle + |011\rangle + |100\rangle - |101\rangle + |110\rangle + |111\rangle \right)|-\rangle
$$

These coefficients are represented graphically by Figure 5.5 and the state $|\psi_0\rangle$ is represented by Figure 5.6.

$|\psi_1\rangle$ can be written as a function of $|A\rangle$ and $|B\rangle = |x^*\rangle$ as follows:

$$
\begin{aligned}
|\psi_1\rangle &= \alpha|A\rangle - \beta|B\rangle \\
&= \alpha|A\rangle - \alpha^*|x^*\rangle \\
&= \frac{1}{2\sqrt{2}}|A\rangle - \frac{\sqrt{7}}{2\sqrt{2}}|x^*\rangle
\end{aligned}
\quad (5.7)
$$

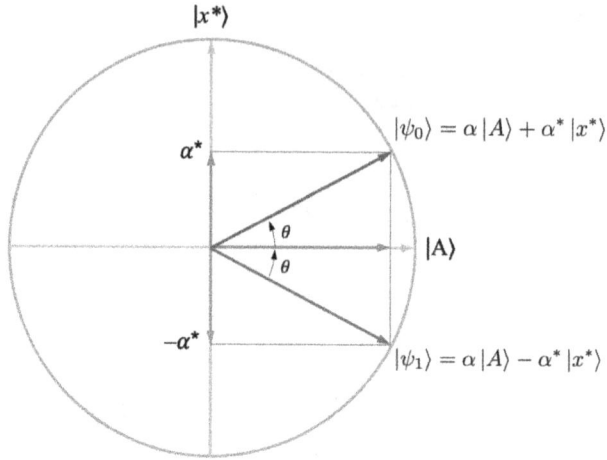

Figure 5.6 Phase inversion state.

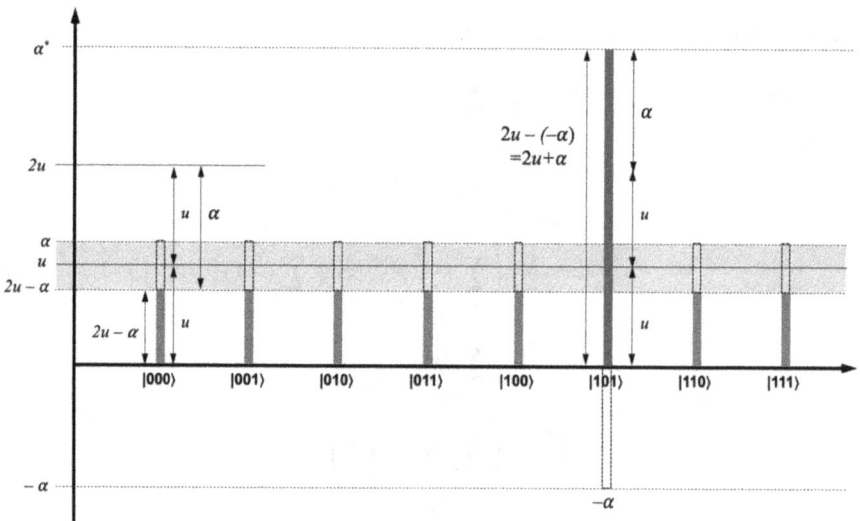

Figure 5.7 The coefficients after the inversion about the mean step.

5.4.3 INVERSION ABOUT THE MEAN

Sometimes referred to as the "Amplitude Amplification", this step involves reversing the coefficients with respect to u, the average of the coefficients of the state $|\psi_1\rangle$ obtained during the previous step. As shown in Figure 5.7, this step leads to reduced values of all the coefficients, except the coefficient α^* which will instead be increased.

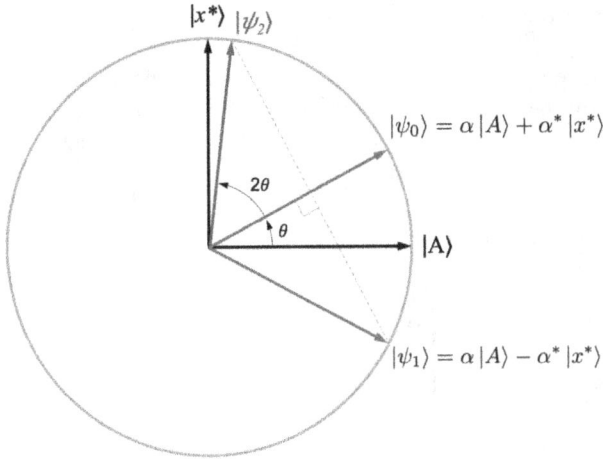

Figure 5.8 The state $|\psi_2\rangle$ after the inversion about the mean.

The new coefficients can be obtained by the transformation:

$$t(\alpha_i) = 2u - \alpha_i \tag{5.8}$$

where, $u = \frac{1}{N}\sum_{i=0}^{N}\alpha_i$, hence,

$$t(\alpha_i) = 2\left(\frac{1}{N}\sum_{j=0}^{N}\alpha_j\right) - \alpha_i \tag{5.9}$$

For example, $\alpha_{i,i=0,\dots,7}$

Since all the coefficients α_i are equal to α except for α^* which is equal to $-\alpha$, then, by applying the transformation given by Equation 5.8, all the coefficients are equal to $(2u - \alpha)$, except for x^* which is equal to $(2u - (-\alpha)) = 2u + \alpha$. This leads to a reduced probability of $|(2u - \alpha)|^2$ to obtain all the states $|x_i\rangle$ except the state x^* for which the probability to obtain it is increased and equal to $|(2u + \alpha)|^2$. Grover's algorithm is based on this step and the previous one (Section 5.4.2) which must be repeated until the state $|\psi_2\rangle$ is closed or equal to the targeted state $|x^*\rangle$. Therefore, the probability to obtain $|x^*\rangle$ is close to 1 and the probability to obtain the other states $|A\rangle$ is close to 0. Figure 5.8 illustrates this transformation.

In the case of our example, the mean u is equal to:

$$u = \frac{1}{8}\left(\sum_{i=0}^{i=6}\frac{1}{2\sqrt{2}} - \frac{1}{2\sqrt{2}}\right) = \frac{1}{8}\left(\sum_{i=0}^{i=5}\frac{1}{2\sqrt{2}}\right) = \frac{3}{8\sqrt{2}}$$

Hence,

$$\alpha_{i,i\neq 5} = 2\frac{3}{8\sqrt{2}} - \frac{1}{2\sqrt{2}} = \frac{3}{4\sqrt{2}} - \frac{2}{4\sqrt{2}} = \frac{1}{4\sqrt{2}}$$

$$\alpha^* = \alpha_5 = 2\frac{3}{8\sqrt{2}} + \frac{1}{2\sqrt{2}} = \frac{3}{4\sqrt{2}} + \frac{2}{4\sqrt{2}} = \frac{5}{4\sqrt{2}}$$

To find the quantum circuit that can carry out this operation, we present another way to perform the inversion about the mean. Instead of inversing the coefficients about the mean, we can do the inversion about the $|0\rangle^{\otimes N} = |000\rangle$. To do this, first, we must transform the superposition state $|\psi_0\rangle$ to the state $|0\rangle^{\otimes N}$, which can be done using the Hadamard transformation. Then, we can do the inversion about the state $|0\rangle^{\otimes N}$ using the following matrix:

$$\begin{pmatrix} 1 & 0 & 0 & \cdots & 0 \\ 0 & -1 & 0 & \cdots & 0 \\ 0 & 0 & -1 & \cdots & 0 \\ \vdots & \vdots & \vdots & \ddots & \vdots \\ 0 & 0 & 0 & \cdots & -1 \end{pmatrix} \tag{5.10}$$

which leads to the following transformation M:

$$M = H^{\otimes n} M H^{\otimes n}$$

$$= H^{\otimes n} \begin{pmatrix} 1 & 0 & 0 & \cdots & 0 \\ 0 & -1 & 0 & \cdots & 0 \\ 0 & 0 & -1 & \cdots & 0 \\ \vdots & \vdots & \vdots & \ddots & \vdots \\ 0 & 0 & 0 & \cdots & -1 \end{pmatrix} H^{\otimes n}$$

$$= H^{\otimes n} \left(\begin{pmatrix} 2 & 0 & 0 & \cdots & 0 \\ 0 & 0 & 0 & \cdots & 0 \\ 0 & 0 & 0 & \cdots & 0 \\ \vdots & \vdots & \vdots & \ddots & \vdots \\ 0 & 0 & 0 & \cdots & 0 \end{pmatrix} - \begin{pmatrix} 1 & 0 & 0 & \cdots & 0 \\ 0 & 1 & 0 & \cdots & 0 \\ 0 & 0 & 1 & \cdots & 0 \\ \vdots & \vdots & \vdots & \ddots & \vdots \\ 0 & 0 & 0 & \cdots & 1 \end{pmatrix} \right) H^{\otimes n} \tag{5.11}$$

$$= H^{\otimes n} \begin{pmatrix} 2 & 0 & 0 & \cdots & 0 \\ 0 & 0 & 0 & \cdots & 0 \\ 0 & 0 & 0 & \cdots & 0 \\ \vdots & \vdots & \vdots & \ddots & \vdots \\ 0 & 0 & 0 & \cdots & 0 \end{pmatrix} H^{\otimes n} - H^{\otimes n} \mathbb{I} H^{\otimes n}$$

$$= H^{\otimes n} \begin{pmatrix} 2 & 0 & 0 & \cdots & 0 \\ 0 & 0 & 0 & \cdots & 0 \\ 0 & 0 & 0 & \cdots & 0 \\ \vdots & \vdots & \vdots & \ddots & \vdots \\ 0 & 0 & 0 & \cdots & 0 \end{pmatrix} H^{\otimes n} - \mathbb{I}$$

$$
= \frac{1}{\sqrt{N}}
\begin{pmatrix}
1 & 1 & 1 & \cdots & 1 \\
1 & & & \cdots & \\
1 & & & \cdots & \\
\vdots & \vdots & \vdots & \ddots & \vdots \\
1 & & & \cdots &
\end{pmatrix}
\begin{pmatrix}
2 & 0 & 0 & \cdots & 0 \\
0 & 0 & 0 & \cdots & 0 \\
0 & 0 & 0 & \cdots & 0 \\
\vdots & \vdots & \vdots & \ddots & \vdots \\
0 & 0 & 0 & \cdots & 0
\end{pmatrix}
H^{\otimes n} - \mathbb{I}
$$

$$
= \frac{1}{\sqrt{N}}
\begin{pmatrix}
2 & 0 & 0 & \cdots & 0 \\
2 & 0 & 0 & \cdots & 0 \\
2 & 0 & 0 & \cdots & 0 \\
\vdots & \vdots & \vdots & \ddots & \vdots \\
2 & 0 & 0 & \cdots & 0
\end{pmatrix}
H^{\otimes n} - \mathbb{I}
\tag{5.12}
$$

$$
= \frac{1}{\sqrt{N}}
\begin{pmatrix}
2 & 0 & 0 & \cdots & 0 \\
2 & 0 & 0 & \cdots & 0 \\
2 & 0 & 0 & \cdots & 0 \\
\vdots & \vdots & \vdots & \ddots & \vdots \\
2 & 0 & 0 & \cdots & 0
\end{pmatrix}
\frac{1}{\sqrt{N}}
\begin{pmatrix}
1 & 1 & 1 & \cdots & 1 \\
1 & & & \cdots & \\
1 & & & \cdots & \\
\vdots & \vdots & \vdots & \ddots & \vdots \\
1 & & & \cdots &
\end{pmatrix}
- \mathbb{I}
$$

$$
= \frac{1}{N}
\begin{pmatrix}
2 & 2 & 2 & \cdots & 2 \\
2 & 2 & 2 & \cdots & 2 \\
2 & 2 & 2 & \cdots & 2 \\
\vdots & \vdots & \vdots & \ddots & \vdots \\
2 & 2 & 2 & \cdots & 2
\end{pmatrix}
- \mathbb{I}
\tag{5.13}
$$

$$
=
\begin{pmatrix}
\frac{2}{N} - 1 & \frac{2}{N} & \frac{2}{N} & \cdots & \frac{2}{N} \\
\frac{2}{N} & \frac{2}{N} - 1 & \frac{2}{N} & \cdots & \frac{2}{N} \\
\frac{2}{N} & \frac{2}{N} & \frac{2}{N} - 1 & \cdots & \frac{2}{N} \\
\vdots & \vdots & \vdots & \ddots & \vdots \\
\frac{2}{N} & \frac{2}{N} & \frac{2}{N} & \cdots & \frac{2}{N} - 1
\end{pmatrix}
$$

Let us calculate the state $|\psi_2\rangle$ for the state $|\alpha_1, \alpha_2, \alpha^*, \cdots, \alpha_N\rangle$:

$$
|\psi_2\rangle = M(\alpha_1, \alpha_2, \alpha^*, \cdots, \alpha_N)
$$

$$
=
\begin{pmatrix}
\frac{2}{N} - 1 & \frac{2}{N} & \frac{2}{N} & \cdots & \frac{2}{N} \\
\frac{2}{N} & \frac{2}{N} - 1 & \frac{2}{N} & \cdots & \frac{2}{N} \\
\frac{2}{N} & \frac{2}{N} & \frac{2}{N} - 1 & \cdots & \frac{2}{N} \\
\vdots & \vdots & \vdots & \ddots & \vdots \\
\frac{2}{N} & \frac{2}{N} & \frac{2}{N} & \cdots & \frac{2}{N} - 1
\end{pmatrix}
\begin{pmatrix}
\alpha_1 \\
\alpha_2 \\
\alpha^* \\
\vdots \\
\alpha_N
\end{pmatrix}
\tag{5.14}
$$

$$
|\psi_2\rangle_3 = \frac{2}{N}\alpha_1 + \frac{2}{N}\alpha_2 + \left(\frac{2}{N} - 1\right)\alpha^* + \cdots + \frac{2}{N}\alpha_N
$$

$$
= 2\left(\frac{1}{N}\sum_{i=1}^{N}\alpha_i\right) - \alpha^*
\tag{5.15}
$$

$$
= 2u - \alpha^*
$$

which is the same result given by Equations (5.8) and (5.9).

If we apply this matrix to our example, we obtain:

$$|\psi_2\rangle = M\left(\frac{1}{4\sqrt{2}}, \frac{1}{4\sqrt{2}}, \frac{1}{4\sqrt{2}}, \frac{1}{4\sqrt{2}}, \frac{1}{4\sqrt{2}}, \frac{5}{4\sqrt{2}}, \frac{1}{4\sqrt{2}}, \frac{1}{4\sqrt{2}}\right)$$

$$= \begin{pmatrix} \frac{2}{8}-1 & \frac{2}{8} & \frac{2}{8} & \cdots & \frac{2}{8} \\ \frac{2}{8} & \frac{2}{8}-1 & \frac{2}{8} & \cdots & \frac{2}{8} \\ \frac{2}{8} & \frac{2}{8} & \frac{2}{8}-1 & \cdots & \frac{2}{8} \\ \vdots & \vdots & \vdots & \ddots & \vdots \\ \frac{2}{8} & \frac{2}{8} & \frac{2}{8} & \cdots & \frac{2}{8}-1 \end{pmatrix} \begin{pmatrix} \frac{1}{4\sqrt{2}} \\ \vdots \\ \frac{1}{4\sqrt{2}} \\ \frac{5}{4\sqrt{2}} \\ \frac{1}{4\sqrt{2}} \\ \frac{1}{4\sqrt{2}} \end{pmatrix} \qquad (5.16)$$

5.4.4 THE MAXIMUM NUMBER OF ITERATIONS

As shown in Figure 5.8, the inversion about the mean leads to a rotation of 2θ of the initial state $|\psi_0\rangle$. Each additional iteration leads to an additional rotation of 2θ. Finally, if we consider the step of Phase inversion, each iteration k leads to a rotation of $\theta + 2k\theta = (1+2k)\theta$.

Now it is possible to calculate the maximum number of the iterations k to reach the final solution which must be close or equal to $|x^*\rangle$. Thus, if $|\psi_2\rangle$ is close or equal to $|x^*\rangle$, then,

$$\begin{aligned} |\psi_2\rangle &= 0|A\rangle + 1|B\rangle \\ &= 1|B\rangle \\ &= \sin((1+2k)\theta)|B\rangle \end{aligned} \qquad (5.17)$$

$$\begin{aligned} \implies \sin((1+2k)\theta) &= 1 \\ &= \sin^{-1} 1 \\ &= \frac{\pi}{2} \end{aligned} \qquad (5.18)$$

$$\implies k = \frac{\pi}{4}\sqrt{\frac{N}{a}} - \frac{1}{2}$$

Therefore,

$$k = \left\lfloor \frac{\pi}{4}\sqrt{N} \right\rfloor \qquad (5.19)$$

We conclude from this last equation that the complexity of Grover's algorithm is $O(\sqrt{N})$. For example, for a password with 80 bits, a maximum $2^{80} = 1.2e24$ iterations are needed to find the password. However, with the quantum algorithm, a maximum $\sqrt{80} = 8$ iterations are needed.

5.5 GROVER'S CIRCUIT

The Grover's circuit is given as follows [1].

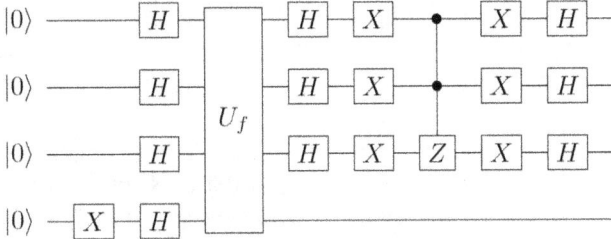

We can simplify the circuit using the following annotations:

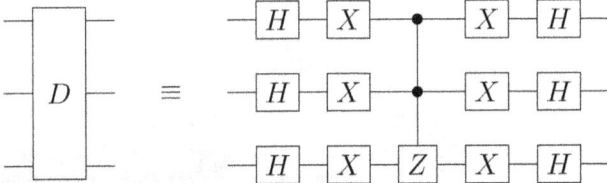

Based on this simplification, we can represent the Grover's circuit with two iterations as follows:

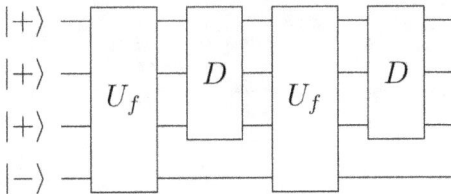

The Grover's circuit with any number of iterations n is given as follows:

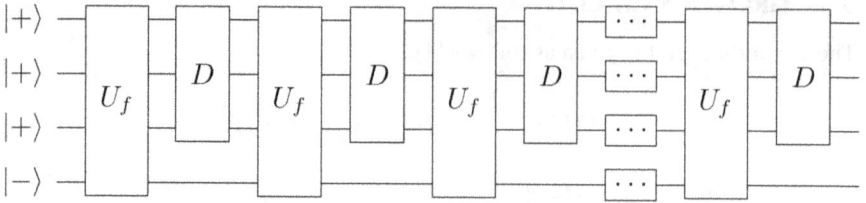

Note that the block D is the one carrying out the inversion about the mean. It is called *Diffuser* or *Amplifier*. The one proposed here is not using the exact matrix presented by Equation (5.10). However, it uses the negative of this matrix, i.e.,

$$\begin{pmatrix} -1 & 0 & 0 & \cdots & 0 \\ 0 & 1 & 0 & \cdots & 0 \\ 0 & 0 & 1 & \cdots & 0 \\ \vdots & \vdots & \vdots & \ddots & \vdots \\ 0 & 0 & 0 & \cdots & 1 \end{pmatrix} \tag{5.20}$$

This leads to negative coefficients α_i instead of positive ones. However, these negative values will not affect the probabilities since their values are equal to $|\alpha_i|^2$ which remains positive and the same. In this case, Figure 5.9 shows how the inversion about the mean step looks for negative coefficients:

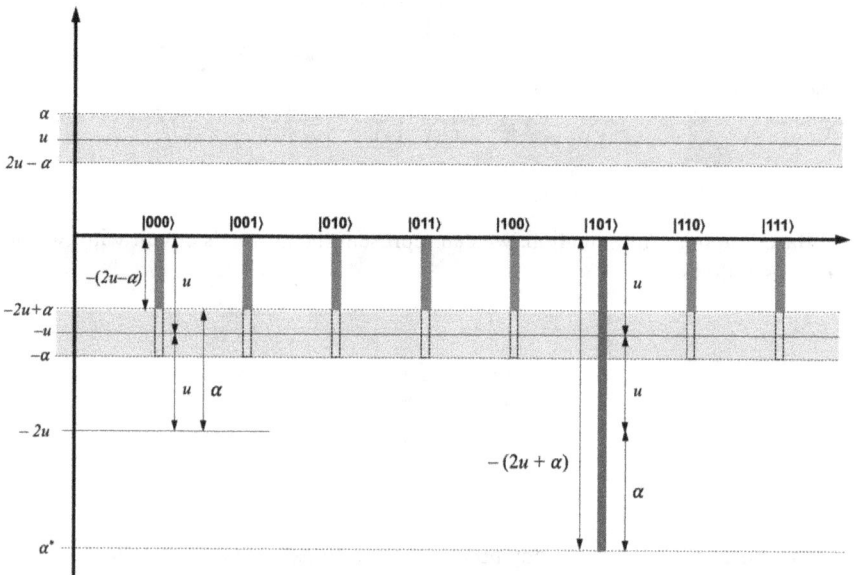

Figure 5.9 The state $|\psi_2\rangle$ after the inversion about the mean with negative values of the coefficients.

5.6 GROVER'S CIRCUIT EXAMPLE

To show how Grover's algorithm is working on an example, consider the following circuit where $|x^*\rangle = |101\rangle$:

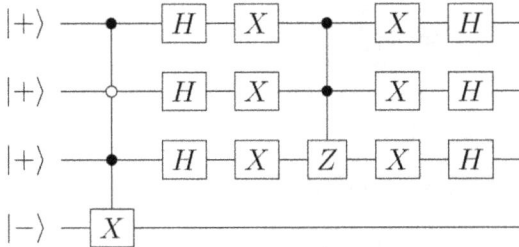

The non-compact version of Grover's algorithm circuit can be represented as follows:

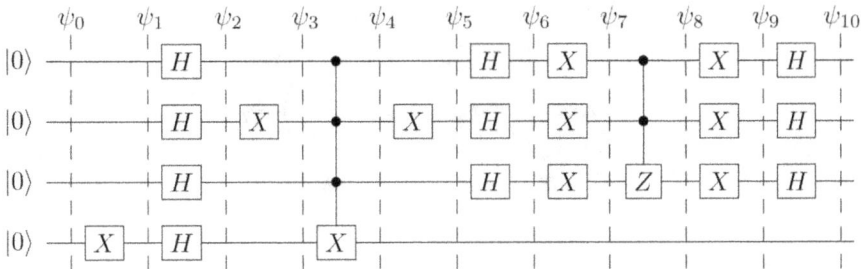

Execute the different steps of this circuit:

$$|\psi_0\rangle = |0\rangle^{\otimes 4} = |0000\rangle = |000\rangle|0\rangle \tag{5.21}$$

$$|\psi_1\rangle = I^{\otimes 3}.X|\psi_0\rangle = I.I.I|000\rangle X|0\rangle = |000\rangle|1\rangle \tag{5.22}$$

$$|\psi_2\rangle = H^{\otimes 4}|\psi_1\rangle = H.H.H.H|\psi_1\rangle = H.H.H|000\rangle H|1\rangle = |+++\rangle|-\rangle$$
$$= \frac{1}{\sqrt{8}}(|000\rangle + |001\rangle + |010\rangle + |011\rangle + |100\rangle + |101\rangle + |110\rangle + |111\rangle)|-\rangle \tag{5.23}$$

$$|\psi_3\rangle = I.X.I.I|\psi_2\rangle = I.X.I|+++\rangle I|-\rangle$$
$$|\psi_3\rangle = \frac{1}{\sqrt{8}}(|010\rangle + |011\rangle + |000\rangle + |001\rangle + |110\rangle + |111\rangle + |100\rangle + |101\rangle)|-\rangle \tag{5.24}$$

We remark here that the state that will generate $|111\rangle$ activating the state CCCX in the next step ($|\psi_4\rangle$) is the state $|101\rangle$.

$$|\psi_4\rangle = CCCX(|\psi_3\rangle)$$
$$|\psi_4\rangle = \frac{1}{\sqrt{8}}(|010\rangle + |011\rangle + |000\rangle + |001\rangle + |110\rangle - |111\rangle + |100\rangle + |101\rangle)|-\rangle \tag{5.25}$$

$$|\psi_5\rangle = I.X.I.I|\psi_4\rangle$$

$$= \frac{1}{\sqrt{8}}(|000\rangle + |001\rangle + |010\rangle + |011\rangle + |100\rangle - |101\rangle + |110\rangle + |111\rangle)|-\rangle$$

$$= \frac{1}{\sqrt{8}}(|000\rangle + |001\rangle + |010\rangle + |011\rangle + |100\rangle + |101\rangle + |110\rangle + |111\rangle - 2|101\rangle)|-\rangle$$

$$= \left(|{+}{+}{+}\rangle - \frac{2}{\sqrt{8}}|101\rangle\right)|-\rangle \tag{5.26}$$

$$|\psi_6\rangle = H^{\otimes 3}|\psi_5\rangle$$

$$= H.H.H\left(|{+}{+}{+}\rangle - \frac{1}{\sqrt{2}}|101\rangle\right)|-\rangle$$

$$= \left(H.H.H|{+}{+}{+}\rangle - \frac{1}{\sqrt{2}}H.H.H|101\rangle\right)|-\rangle \tag{5.27}$$

$$= \left(|000\rangle - \frac{1}{\sqrt{2}}|{-}{+}{-}\rangle\right)|-\rangle$$

$$|\psi_7\rangle = X^{\otimes 3}|\psi_6\rangle$$

$$= X.X.X\left(|000\rangle - \frac{1}{\sqrt{2}}|{-}{+}{-}\rangle\right)|-\rangle$$

$$= \left(X.X.X|000\rangle - \frac{1}{\sqrt{2}}X.X.X|{-}{+}{-}\rangle\right)|-\rangle \tag{5.28}$$

$$= \left(|111\rangle - \frac{1}{\sqrt{2}}|{-}{+}{-}\rangle\right)|-\rangle$$

$$|\psi_8\rangle = CCZ(|\psi_7\rangle)$$

$$= CCZ\left(|111\rangle - \frac{1}{\sqrt{2}}|{-}{+}{-}\rangle\right)|-\rangle$$

$$= \left(CCZ|111\rangle - \frac{1}{\sqrt{2}}CCZ|{-}{+}{-}\rangle\right)|-\rangle$$

$$= \left(-|111\rangle - \frac{1}{\sqrt{2}}\left(|{-}{+}{-}\rangle - \frac{1}{\sqrt{2}}|111\rangle\right)\right)|-\rangle \tag{5.29}$$

$$= \left(-|111\rangle - \frac{1}{\sqrt{2}}|{-}{+}{-}\rangle + \frac{1}{2}|111\rangle\right)|-\rangle$$

$$= \left(\left(\frac{1}{2} - 1\right)|111\rangle - \frac{1}{\sqrt{2}}|{-}{+}{-}\rangle\right)|-\rangle$$

$$|\psi_9\rangle = X^{\otimes 3}(|\psi_8\rangle)$$

$$= X.X.X\left(\left(\frac{1}{2} - 1\right)|111\rangle - \frac{1}{\sqrt{2}}|{-}{+}{-}\rangle\right)|-\rangle$$

$$= \left(-\frac{1}{2}X.X.X|111\rangle - \frac{1}{\sqrt{2}}X.X.X|{-}{+}{-}\rangle\right)|-\rangle \tag{5.30}$$

$$= \left(-\frac{1}{2}|000\rangle - \frac{1}{\sqrt{2}}|{-}{+}{-}\rangle\right)|-\rangle$$

$$|\psi_{10}\rangle = H^{\otimes 3}(|\psi_9\rangle)$$

$$= \left(-\frac{1}{2}H.H.H|000\rangle - \frac{1}{\sqrt{2}}H.H.H|-+-\rangle\right)|-\rangle$$

$$= \left(-\frac{1}{2}|+++\rangle - \frac{1}{\sqrt{2}}|101\rangle\right)|-\rangle$$

$$= \left(-\frac{1}{2}\left(\frac{1}{\sqrt{8}}\sum_{x\in\{0,1\}^3}x\right) - \frac{1}{\sqrt{2}}|101\rangle\right)|-\rangle$$

$$= \left(-\frac{1}{2\sqrt{8}}(|000\rangle + |001\rangle + |010\rangle + |011\rangle + |100\rangle + |101\rangle + |110\rangle + |111\rangle) - \frac{1}{\sqrt{2}}|101\rangle\right)|-\rangle$$

$$= \left(-\frac{1}{2\sqrt{8}}(|000\rangle + |001\rangle + |010\rangle + |011\rangle + |100\rangle + |110\rangle + |111\rangle) - \frac{1}{2\sqrt{8}}|101\rangle - \frac{1}{\sqrt{2}}|101\rangle\right)|-\rangle$$

$$= \left(-\frac{1}{4\sqrt{2}}(|000\rangle + |001\rangle + |010\rangle + |011\rangle + |100\rangle + |110\rangle + |111\rangle) - \frac{5}{4\sqrt{2}}|101\rangle\right)|-\rangle$$

$$(5.31)$$

This means that all the states have the probability of $\left(-\frac{1}{4\sqrt{2}}\right)^2 = \frac{1}{32} \approx 0.03125$, except the state $|101\rangle$, which has the probability $\left(-\frac{5}{4\sqrt{2}}\right)^2 = \frac{25}{32} \approx 0.78125$. Therefore, the code we are searching for is $|101\rangle$. The condition of normalisation is verified:

$$\frac{1}{32} + \frac{1}{32} + \frac{1}{32} + \frac{1}{32} + \frac{1}{32} + \frac{1}{32} + \frac{1}{32} + \frac{25}{32} = \frac{7}{32} + \frac{25}{32} = \frac{32}{32} = 1$$

5.7 GROVER'S QISKIT PROGRAM

This section will be devoted to the presentation of the Qiskit program of the Grover's algorithm.

5.7.1 GROVER'S CIRCUIT FOR $N = 3$

The Qiskit program of the Grover's circuit with $n = 3$ and $|x^*\rangle = |101\rangle$. The number of shots, i.e., the number of times the measures are taken is set to 1000 in line 24.

```
1   import qiskit as q
2   simulator = q.Aer.get_backend('qasm_simulator')
3   circuit = q.QuantumCircuit(4,3)
4   # H.H.H
5   circuit.h([0,1,2])
6   # |->
7   circuit.x(3)
8   circuit.h(3)
9   for i in range(2):
10      # Uf: f(101)=1
11      circuit.x(1)
12      circuit.mct([0,1,2],3)
13      circuit.x(1)
```

```
14    circuit.h([0,1,2]) # H.H.H
15    circuit.x([0,1,2]) # X.X.X
16    # CCZ
17    circuit.h(2)
18    circuit.ccx(0,1,2)
19    circuit.h(2)
20    circuit.x([0,1,2]) # X.X.X
21    circuit.h([0,1,2]) # H.H.H
22  circuit.measure([0,1,2], [0,1,2])
23  print(circuit.draw(output='text'))
24  job = q.execute(circuit, simulator, shots=1000)
25  result = job.result()
26  output = result.get_counts(circuit)
27  print(output)
28  import matplotlib.pyplot as plt
29  q.visualization.plot_histogram(output)
30  plt.show()
```

The execution of this program leads to the following graph where we see that the probability to get the state $|101\rangle$ is $953/1000 = 95\%$ and the probability to get the other states is ≈ 0.

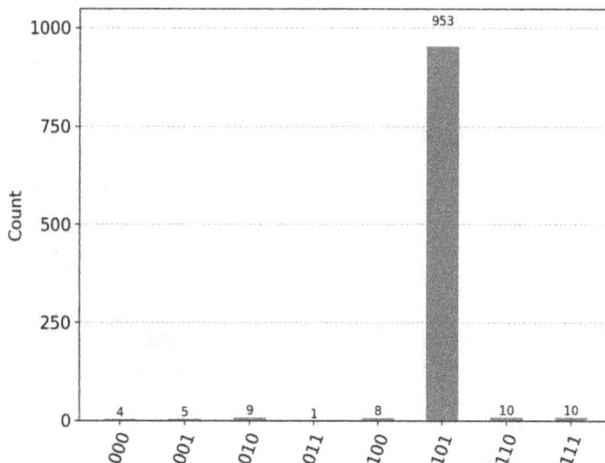

Figure 5.10 Histogram output of the Grover program for $|x^*\rangle = |101\rangle$.

5.7.2 GROVER'S CIRCUIT FOR ANY VALUE OF N

The Qiskit program of the Grover's circuit with any value of n is given by the following program. In this program, the value of n is fixed to 4 in line 4 and the circuit of x^* is set to $|11...1\rangle$ in line 13.

```
import qiskit as q
import math
simulator = q.Aer.get_backend('qasm_simulator')
n = 4
circuit = q.QuantumCircuit(n+1,n)
circuit.x(n)
#Hx(n+1)
circuit.h(range(n+1))
#Number of iterations K
K = int(math.pi/4.*math.sqrt(2**n))
for j in range(K):
    #Uf: f(11...1)=1
    circuit.mct(list(range(n)),n)
    #HX
    circuit.h(range(n+1))
    circuit.x(range(n+1))
    #MCZ
    circuit.h(n-1)
    circuit.mct(list(range(n-1)),n-1)
    circuit.h(n-1)
    #XH
    circuit.x(range(n+1))
    circuit.h(range(n+1))
circuit.measure(range(n), range(n))
print(circuit.draw(output='text'))
job = q.execute(circuit, simulator, shots=1000)
result = job.result()
output = result.get_counts(circuit)
print(output)
import matplotlib.pyplot as plt
q.visualization.plot_histogram(output)
plt.show()
```

5.8 CONCLUSION

In this chapter, we explored Grover's algorithm, a cornerstone of quantum computing, known for its efficiency in searching unsorted databases. The chapter explained the theoretical constructs of Grover's algorithm, focusing on its potential to significantly reduce the computational resources needed for brute-force search tasks compared to classical approaches. Through a series of detailed explanations and practical examples using the Qiskit framework, the chapter demonstrated the algorithm's effectiveness and practicality.

Grover's algorithm utilises quantum superposition and entanglement to create a state that includes all possible answers. By iteratively applying quantum interference, the probability amplitude of the correct answer is amplified, allowing it to be observed with a higher probability upon measurement. This process, known as amplitude amplification, is a powerful demonstration of quantum speed-up, providing a quadratic improvement over any classical algorithm.

The practical sections of this chapter showed how Grover's algorithm could be implemented for specific instances, including the setup and execution of the quantum circuits within the Qiskit programming environment. These examples not only highlighted the algorithm's adaptability across different scenarios but also showcased its scalability, which is crucial for its application in real-world problems.

As quantum computing continues to evolve, the insights provided by Grover's Algorithm will undoubtedly play a critical role in the development of new quantum technologies. Its applications extend beyond simple database searches, offering potential advances in fields ranging from cryptography to optimisation and beyond.

In conclusion, Grover's algorithm represents a paradigm shift in how we approach problem-solving with quantum computers. Its ability to efficiently search through unsorted data using fewer resources than classical computers could lead to significant advancements in various technological areas. As researchers continue to refine quantum algorithms and hardware, the principles and techniques outlined in this chapter are essential for harnessing the full potential of quantum computing.

REFERENCES

1. Grover, L.K.: A fast quantum mechanical algorithm for database search. Proceedings of the Twenty-Eighth Annual ACM Symposium on Theory of Computing – STOC '96, pp. 212–219. Association for Computing Machinery, Philadelphia, PA (1996). https://doi.org/10.1145/237814.237866

2. Baaquie, B.E., Kwek, L.-C.: Grover's Algorithm, pp. 185–202. Springer, Singapore (2023). https://doi.org/10.1007/978-981-19-7517-2_11

3. LaPierre, R.: Grover Algorithm, pp. 163–176. Springer, Cham (2021). https://doi.org/10.1007/978-3-030-69318-3_12

4. Kun, D., Strömberg, T., Spagnolo, M., Dakić, B., Rozema, L.A., Walther, P.: Direct and Efficient Detection of Quantum Superposition (2024). https://arxiv.org/abs/2405.08065

5. Markidis, S.: What is Quantum Parallelism, Anyhow? (2024). https://arxiv.org/abs/2405.07222

6 Quantum Key Distribution Principles and Protocols

Mohammad Hammoudeh, Clinton M. Firth,
Ahcene Bounceur, Bamidele Adebisi,
and Danish Vasan

6.1 INTRODUCTION TO QUANTUM KEY DISTRIBUTION

Quantum Key Distribution (QKD) is a secure communication method that uses the principles of quantum mechanics to enable two parties to establish a shared random secret key. This symmetric key is known only to the communicating parties and is used to ensure confidentiality and integrity in communication channels. Unlike traditional encryption methods, which rely on mathematical complexity, QKD relies on the fundamental principles of quantum mechanics, such as the entangled particles, Heisenberg's uncertainty principle, and the no-cloning theorem. These principles ensure that any attempt to eavesdrop on the communication channel will disturb the quantum state, immediately revealing to the legitimate parties the presence of the intruder or error in the communication channel. Through the exchange of quantum states, usually encoded in photons, the sender and receiver can establish a shared key. The properties of quantum mechanics make the key resistant to interception or tampering attempts. The security of QKD is based on fundamental physical laws rather than mathematical or computational complexity.

As quantum computing technology advances, traditional encryption methods that rely on mathematical algorithms, such as RSA and ECC, become more vulnerable to decryption via quantum algorithms, like Shor's algorithm. Quantum computers demonstrated the potential to break current cryptographic algorithms, putting sensitive data at risk. QKD addresses this threat by leveraging the principles of quantum mechanics to create secure communication channels immune to quantum attacks. Unlike traditional encryption methods, QKD offers a method for distributing encryption keys with unconditional security guarantees. Thus, the need for QKD arises from the need to protect sensitive data and communications against the imminent threat posed by quantum computing capabilities.

The emergency of QKD stems from the imperative to secure communications amidst the rising susceptibility of classical cryptographic systems to attacks that are facilitated by quantum computers. The roots of QKD trace back to the early 20th

DOI: 10.1201/9781003475286-6

century, with the development of quantum mechanics and cryptography. Groundbreaking work by physicists such as Max Planck, Albert Einstein, Niels Bohr, and others laid the foundation for understanding the quantum nature of particles and their behaviour. The concept of using quantum principles for cryptography was first introduced in the 1970s [1]. In 1970, Stephen Wiesner presented the concept of quantum money and quantum conjugate coding, which established the basis for quantum cryptography. However, his work remained mostly theoretical at the time. The innovation in practical QKD was in 1984 with the proposal of the BB84 protocol by Charles Bennett and Gilles Brassard [2].

In the BB84 protocol, one communicating party prepares a series of quantum states, each representing a bit of the secret key, and sends them to the other party over a quantum communication channel. The receiver measures each received state using one of two complementary bases. After exchanging classical information over a public channel to compare measurement bases and discard measurements affected by the eavesdropper, the two communicating parties are left with a subset of shared bits that form their secret key. Following the introduction of the BB84 protocol, experimental demonstrations of QKD appeared in the early 1990s. Researchers including Artur Ekert, Bennett, and others conducted revolutionising experiments to implement QKD protocols in laboratory settings, demonstrating the feasibility of quantum-based secure communication. Over the following decades, QKD saw remarkable advancements and improvements in both theoretical understanding and practical implementation. New protocols were presented, and experimental techniques were developed, resulting in higher key generation rates, longer transmission distances, and enhanced security. In the early 2000s, QKD began to evolve from academic research to commercial applications. Several companies began offering QKD protocols for secure communication, targeting critical infrastructure sectors such as defence, government, finance, and healthcare. This indicated an important landmark in the commercialisation and adoption of quantum cryptography for real-world applications.

Today, QKD remains an active area of research and development, with continuing efforts to resolve technical challenges, enhance performance, and investigate novel applications. It holds the promise to enhance the security of communication systems amidst the rising threat posed by quantum computers. As quantum computers have the potential to break conventional cryptographic systems, QKD becomes a suitable alternative for secure key distribution. QKD offers unconditional security guarantees based on the laws of physics, making it immune to classical cryptographic attacks, such as brute-force or mathematical attacks. However, practical implementations of QKD face several challenges, including limited transmission distances, channel noise, infrastructure requirements, and vulnerability to certain types of side-channel attacks. Despite these challenges, QKD has the potential to enable secure communication in future quantum networks. This chapter presents the concept of QKD and its applications. It examines various studies and protocols by researchers in the field of quantum information science and cryptography.

6.2 PRINCIPLES OF QUANTUM MECHANICS RELEVANT TO QKD

In this section, we review the basic principles of quantum mechanics that underline QKD protocols. Knowledge of these principles is necessary for understanding the mechanisms underpinning secure key exchange in quantum communication. By briefly reviewing these fundamental principles in quantum mechanics, this section provides readers with the necessary background to understand the details of QKD protocols discussed later in the chapter. Understanding the concepts governing quantum communications is essential for designing, implementing, and analysing secure communication protocols based on quantum mechanics.

6.2.1 QUANTUM SUPERPOSITION

Quantum superposition permits quantum particles to exist in multiple states simultaneously until measured or observed. This phenomenon forms the basis of many QKD protocols. It allows the encoding of information onto quantum states for secure transmission. In conventional physics, objects are typically defined as being in a single definite state at any given time. However, in quantum mechanics, particles can exist in a state that is a linear combination or summation of several possible states. For instance, an electron can be in a superposition of both spin-up and spin-down states simultaneously until it is observed.

The superposition principle permits quantum objects to exist in a state that is a combination of different possibilities, with each possibility having a probability amplitude. When a measurement or observation is made on the system, it "collapses" into one of the possible states. The probability of finding the system in a particular state is associated with the square of the amplitude of that state. Quantum superposition is one of the main properties that distinguishes quantum mechanics from classical physics. It performs an essential function in phenomena such as quantum entanglement and quantum computing, providing means for more powerful and innovative applications in various fields.

6.2.2 QUANTUM ENTANGLEMENT

Quantum entanglement is a unique feature of quantum physics that describes the unbreakable correlation between the quantum states of entangled particles regardless of the distances between them. Entanglement is a central aspect of quantum mechanics, which enables its potential applications in various fields, including cryptography and quantum computing. In quantum theory, states are defined by mathematical objects called wave functions.

Albert Einstein, Boris Podolsky, and Nathan Rosen (EPR) [3] presented the concept of quantum entanglement, showing how measurements on entangled quantum systems can result in seemingly paradoxical results. They described an EPR pair, consisting of two particles with complementary properties. When measurements are made on both particles, their states are found to be correlated. This phenomenon, termed "spooky action at a distance" by Einstein, suggested the possibility of faster-than-light communication. However, upon closer examination, the apparent paradox dissolves. The correlations between entangled particles are an effect of quantum mechanics and do not violate the laws of physics. The EPR effect ultimately highlights the fundamental complementarity of quantum systems and the non-local correlations that can exist between them.

Daniel Greenberger, Michael Horne, and Anton Zeilinger discovered a significant demonstration of quantum entanglement involving three entangled quantum particles, known as q-ons, prepared in a special state called the GHZ state [4]. In their experiment, three distant experimenters independently choose to measure either the shape or colour of their respective q-ons, recording the results. Square shapes and red colours are called "good", and circular shapes and blue colours are called "evil". Upon comparing their measurements, they found a surprising pattern: when two experimenters measured the shape and the third measured colour, they consistently found either 0 or 2 "evil" results (circular shape or blue colour), whereas when all three measured colours, they find either 1 or 3 "evil" results. This outcome, predicted by quantum mechanics and observed experimentally, defies the notion of definite properties in physical systems independent of measurement. The GHZ effect illustrated the inherent limitations of quantum systems and called for a revaluation of traditional assumptions about physical reality.

Entanglement was also discussed beyond multiple quantum particles to the temporal evolution of a single quantum particle. "Entangled histories" arise in situations where it becomes impossible to assign a definite state to a system at every moment in time. When measurements are made to collect partial information about the system's evolution, the system's state cannot be definitively determined at any given time. Entangled histories illustrate the complex and non-intuitive nature of quantum systems, challenging traditional notions of determinism and causality in physics. This quantum temporal equivalent reflects the intuition underlying the Many-Worlds Interpretation (MWI) of quantum mechanics, where definite states turn into mutually contradictory historical trajectories. MWI is a philosophical viewpoint within quantum mechanics that advocates the objective reality of the universal wavefunction and rejects the notion of wavefunction collapse [5]. According to MWI, all possible outcomes of quantum measurements are realised in separate "worlds" or universes. This interpretation, also known as the relative state formulation or the Everett interpretation, was first proposed by physicist Hugh Everett in 1957 [6] and popularised by Bryce DeWitt in the 1970 [7]. In the context of QKD, this means that any attempt to eavesdrop or intercept quantum communication would lead to the creation of multiple parallel universes, each representing a different outcome of the measurement. As a result, any attempt to gain information about the quantum key would disturb the system and alert the legitimate parties to the presence of an eavesdropper. This feature of MWI can potentially enhance the security of QKD protocols by providing a means to detect and prevent unauthorised access to quantum information.

In modern versions of MWI, the collapse of the wave function is explained through quantum decoherence, a mechanism explored since the 1970s. MWI is considered a mainstream interpretation of quantum mechanics alongside other decoherence interpretations, the Copenhagen interpretation, and hidden variable theories like Bohmian mechanics. MWI suggests the existence of an uncountable number of universes, forming a many-branched tree of time where every possible quantum outcome is actualised. This interpretation aims to resolve issues such as the measurement problem and paradoxes like Wigner's friend, the EPR paradox, and Schrödinger's cat, by positing the existence of all possible outcomes in distinct universes.

MWI offers insights into the nature of quantum communication protocols, such as quantum cryptography. In quantum cryptography, MWI provides a conceptual framework for understanding the secure exchange of cryptographic keys using quantum states. By exploiting the principles of quantum superposition and entanglement, MWI offers new possibilities for the development of secure and efficient communication protocols that leverage the unique properties of quantum systems. By considering the existence of multiple parallel universes, MWI offers novel approaches to enhancing the security and efficiency of quantum communication technologies.

6.2.3 NO-CLONING THEOREM

The no-cloning theorem states that perfectly copying an arbitrary unknown quantum state is impossible. This theorem is fundamental for QKD, as it ensures that any attempt by an eavesdropper to intercept and copy quantum states sent between communicating parties will unavoidably introduce errors, alerting legitimate users to potential security breaches. This theorem, first proposed by physicists Wootters and Zurek [8], serves as a basis of quantum theory, distinguishing it from classical physics.

To illustrate the implications of the no-cloning theorem in the context of QKD, consider a scenario where a user sends messages encoded in the form of photons, whose quantum states represent bits. The security rests in the randomness and uncertainty associated with these states. Even if an intruder intercepts a photon, measuring its state collapses it into a definite 0 or 1, which introduces errors that are detectable by both the sender and the receiver. The communicating parties publicly compare a portion of their keys to verify no tampering occurred, confirming a secure communication channel even in the presence of a malicious party. The no-cloning theorem plays a critical role in ensuring the integrity of QKD by preventing attempts at unauthorised key copying or eavesdropping.

6.2.4 UNCERTAINTY PRINCIPLE

The uncertainty principle, first proposed by Werner Heisenberg [9–11], emphasises the basic limit to the precision with which certain pairs of physical properties, such as position and momentum, can be simultaneously measured. This principle plays a role in QKD by enforcing limits on the accuracy of measurements performed on quantum states, contributing to the security of the key distribution process.

In QKD, the uncertainty influences the process of encoding and decoding cryptographic keys using quantum states. One of the key components of QKD protocols, e.g., BB84, is the transmission of quantum states, normally photons, between two parties. These quantum states carry the information used to generate the cryptographic key. The uncertainty principle makes it impossible to accurately measure certain properties of these quantum states. For instance, if a communicating entity attempts to accurately measure the polarisation of a photon, it will introduce uncertainty in the measurement of other complementary properties, such as the photon's phase or amplitude. This makes it difficult for an eavesdropper to intercept and decode the quantum states without introducing detectable errors. Furthermore, the uncertainty principle restricts the amount of information that can be extracted from the quantum states without disturbing them. If an intruder attempts to learn information about the transmitted quantum states, he may inadvertently alter the states, resulting in discrepancies that can be detected by legitimate users. Hence, the uncertainty principle is a fundamental aspect of the security of QKD protocols, contributing to their resilience against eavesdropping and unauthorised access. By leveraging the inherent uncertainty of quantum states, QKD protocols ensure the confidentiality and integrity of cryptographic keys.

6.2.5 QUANTUM STATES AND MEASUREMENT

Quantum states, such as the polarisation of photons or the spin of particles, serve as carriers of information in QKD. The choice of quantum states and measurement bases is fundamental for the security of QKD protocols as it governs the encoding and decoding of information. QKD protocols use different sets of quantum states and measurement bases to ensure secure key exchange. Data encoding in quantum communications converts classical information into quantum states, typically in the form of qubits, for transmission over quantum channels. Different protocols use different encodings, such as polarisation (horizontal or vertical) or phase shift, for representing 0 and 1 in these superposed states. Several processes are used for encoding classical bits of information into quantum states:

1. Basis Encoding Choice: In QKD protocols like BB84, classical bits are encoded into quantum states based on specific quantum properties, such as polarisation or phase. For example, a classical bit 1 may be encoded as a horizontal polarisation state and a classical bit 1 as a vertical polarisation state. The sender randomly selects a measurement basis, e.g., polarisation or phase, to encode each qubit. The receiver needs to know this basis to interpret the received qubits.
2. Quantum States Superposition: Quantum information can also be encoded using the superposition principle, where qubits are made in a linear combination of various quantum states. This enables encoding multiple classical bits into one qubit, which improves the efficiency of quantum communication.
3. Entanglement: Quantum entanglement allows the encoding of information of multiple qubits in a correlated manner. This means that changes to one qubit can immediately change its entangled partner, enabling secure transmission of data over long distances.
4. Measurement Outcome: When a receiver measures a received qubit, the superposition "collapses" into a definite 0 or 1 based on the chosen measurement basis. This measurement operation is vital for extracting the encoded information.

5. Error Detection: The sender and receiver publicly share a portion of their bases after transmission for error detection purposes. Any attempt to intercept and measure the qubits introduces errors. Due to the inherent randomness and correlated bases, the sender and receiver can statistically detect discrepancies in their shared key, revealing tampering.

6. Quantum Error Correction Codes: Quantum error correction codes are designed to protect transmitted data from errors caused by noise or imperfections in the quantum channel. These codes involve encoding classical information into quantum states in a way that errors can be detected and corrected during the decoding process.

Data encoding in quantum communications is essential for ensuring the security, reliability, and efficiency of QKD protocols. The security of QKD protocols depends on the impossibility of copying an unknown quantum state (no-cloning theorem). Any malicious attempt to measure or copy qubits certainly disturbs them, introducing detectable errors. The randomness of superposition and the correlation due to entanglement further strengthen QKD's protocols resistant to eavesdropping and noise by making it impossible for an intruder to predict or manipulate the qubits without being detected.

6.3 QUANTUM KEY DISTRIBUTION PROTOCOLS

QKD protocols form the foundations of secure communication operating on the principles of quantum mechanics. This section provides an overview of some of the most prominent QKD protocols in the literature to date. These protocols vary in their approach, complexity, and security guarantees, but all share the common objective of securely exchanging cryptographic keys between two parties.

There are other published QKD protocols, and they continue to either enhance BB84 or other protocols to address their deficiencies. QKD is an ongoing field of research there are many such protocols, but for completeness and to highlight a few that will not be covered in this book for readers to further educate themselves: BBM92, B92, COW, DPS, HDQKD, KMB09, MSZ96, SARG04, Six-state, three-stage quantum cryptography, and T12.

6.3.1 BB84 PROTOCOL

The BB84 protocol, named after its inventors Charles Bennett and Gilles Brassard, is one of the earliest QKD protocols. It was developed in 1984 and serves as a primary technique for securely exchanging cryptographic keys between two parties to achieve unbreakable encryption. BB84 relies on the properties of quantum mechanics to communicate data securely over quantum channels. Such quantum principles allow two communicating parties, often referred to as Alice (the sender) and Bob (the receiver) to create a shared secret key without the risk of interception. Alice encodes the transmission of photons with quantum states, e.g., polarisation, and Bob makes subsequent measurements of these states. These particles serve as carriers of information. Through randomly taking measurement bases, Alice and Bob can establish a shared secret key through the transmission of quantum bits (qubits) while detecting any eavesdropping attempts. The protocol comprises respective steps to safeguard the security and integrity of the communication. A simplified explanation of how the BB84 protocol works is given below.

1. Preparation: Alice generates a random stream of photons in one of four possible quantum states, namely horizontal (H), vertical (V), diagonal (D), or anti-diagonal (A). These states represent two mutually unbiased bases, commonly denoted as the rectilinear $(|0\rangle, |1\rangle)$ and diagonal $(|+\rangle, |-\rangle)$ bases.
2. Transmission: Alice then sends the prepared encoded photons to Bob over a quantum communication channel, e.g., free-space satellite or optical fibre communication. As the polarisation basis of each photon is selected randomly by Alice, she does not reveal the basis used for encoding each photon during transmission.
3. Measurement: Bob receives the photons and randomly measures one of the two bases to measure each photon's quantum state, i.e., in either the vertical-horizontal basis or the diagonal-antidiagonal basis. Note that Bob's selection of measurement basis is independent of Alice's selection during the preparation step.
4. Public Announcement or Key Reconciliation: After the transmission, Alice and Bob publicly disclose the bases they used to encode and measure each photon. However, they do not disclose the actual measurement results or the specific polarisations. Also, they discard the qubits measured in different bases. They only hold the bits for which they used the same bases. These bits form the raw key.
5. Comparison and Information Extraction: Alice and Bob compare a subset of their raw key results to check for any errors caused by potential eavesdroppers or channel noise. Through standard error correction and privacy amplification techniques, they discard the photons where their measurement bases did not match and keep the remaining ones, i.e., Alice and Bob reconcile and extract a shared secret key from the remaining qubits that have matched measurement results. This secret key is a shorter but a more secure shared key. Also, Alice and Bob publicly communicate the positions of the qubits used for the error check without revealing their values. The resulting shared key is suitable for use in classical symmetric encryption algorithms.

QKD is a primary concept in the BB84 protocol. It offers unconditional security guarantees against eavesdropping attacks. The key is established based on the quantum states of particles, making it impossible for an eavesdropper to acquire the key without disturbing the quantum states. The BB84 protocol delivers a high level of security due to its dependence on the properties of quantum states, such as the no-cloning theorem and the uncertainty principle. As a reminder, the no-cloning theorem states that it is impossible to generate an identical copy of an unknown quantum state. This property guarantees that any attempt to intercept and replicate the transmitted quantum states will introduce errors, which Alice and Bob can detect during the comparison step. The uncertainty principle, on the other hand, ensures that any attempt to measure the quantum state of a photon will disturb its original state. This property makes it impossible for an eavesdropper to gather complete knowledge of the transmitted quantum states without being detected.

The BB84 protocol has many advantages compared to traditional cryptographic methods. First, BB84 offers unconditional security as the security of the communication is guaranteed by the laws of physics. As long as the laws of quantum mechanics remain valid, the protocol remains unbreakable. Second, the BB84 protocol is resistant to attacks from quantum computers. Classical cryptographic methods, e.g., RSA, can be easily broken by quantum computers using Shor's algorithm. On the contrary, the security of the BB84 protocol relies on the principles of quantum mechanics, rendering it impervious to attacks from quantum computers. Finally, the BB84 protocol implements an efficient key distribution making it practical for real-world applications. The shared secret key can be created with a reasonably small number of exchanged photons.

However, the BB84 protocol has some limitations. It is exposed to photon loss and noise in the quantum channel. These issues can cause errors in the measurement results, which impacts the security of the shared key. Techniques such as error correction and privacy amplification can thwart these issues to a certain degree. There are further practical challenges of the BB84 protocol that affect its implementation in real-world scenarios. Many factors may impact the protocol's performance, including the stability of the quantum channel, the efficacy of photon detectors, and the reliability of quantum state preparation. Additionally, to use the BB84 protocol, it is essential to have a quantum channel for transmitting the quantum states. However, this requirement presents practical challenges related to maintaining a secure quantum channel over long distances. Factors such as decoherence and loss of quantum information in the transmission medium contribute to these challenges. Although methods for safely completing key reconciliation exist, they increase computational costs and might decrease the effectiveness of the key creation process. Consequently, the implementation of BB84 in real-world communication networks becomes a complex task. Another practical implementation challenge of BB84 is the need for a QKD infrastructure that offers robust quantum key distribution hardware and protocols. Setting up and maintaining such infrastructure can be a complex and expensive task, limiting the scalability of BB84 for large-scale adoption. The limited compatibility of BB84 with standard cryptographic systems adds further complexity to its practical utilisation. Although BB84 can be used to create secure keys for symmetric encryption algorithms, integrating it into communication systems may involve significant modifications or the development of new protocols. This constraint makes it difficult to apply BB84 in real-world communication networks as it adds complexity to the overall system. BB84's capacity to produce secure keys at a given rate is constrained by variables like the quantum channel's efficiency and the quantum state transmission error rate. When compared to traditional key exchange techniques, this may lead to comparatively slow key generation rates, which might not be appropriate for high-speed communication networks. Another known limitation of BB84 is its vulnerability to the man in the middle attack and in particular the act of intercept resend, where an eavesdropper intercepts the quantum states sent by

Alice, measures them, and then resends them to the destination, Bob, without any manipulation. While BB84 incorporates mechanisms for detecting intercept-resend attacks, it does not fully mitigate the threat.

BB84 was evaluated and implemented in a variety of experimental configurations. Considerable advancements were achieved by organisations and researchers globally in their efforts to establish the viability of quantum key distribution via the BB84 protocol. When considering applications, the BB84 protocol shows significant potential for facilitating secure communication across a multitude of domains, such as secure data transmission and messaging. For instance, a quantum solution for securing financial transactions against fraud and unauthorised access can be provided by the BB84 protocol. In general, BB84 is suitable for establishing secure communication channels for messaging platforms ensuring the confidentiality and integrity of information. Similarly, BB84 can be applied by highly resourced entities in sensitive critical infrastructure such as communication within the government and military.

6.3.2 E91 PROTOCOL

Although it is among the most extensively studied and well-known quantum cryptographic protocols, the BB84 is not the only one. A multitude of alternative quantum cryptographic protocols were developed to tackle various aspects of secure communication. One such protocol is E91 [12]. The origins of the E91 protocol have roots in the seminal contributions of Boris Podolsky, Nathan Rosen, and Albert Einstein in 1935. In their EPR paper, they introduced the notion of entanglement and prompted research on the exhaustiveness of quantum mechanics. However, Artur Ekert did not propose the E91 Protocol until 1991 as a method for utilizing quantum entanglement to enable secure communication. Ekert's protocol described a secure method for two individuals to exchange a secret key, despite the existence of a potential eavesdropper. The E91 protocol has been subjected to extensive testing and verification since its inception, thereby establishing itself as a fundamental pillar of quantum cryptography.

By utilising the phenomenon of quantum entanglement, the E91 protocol generates a secret key that is shared between two parties. In contrast to the BB84 protocol, which is predicated on the characteristics of individual photons, the E91 protocol capitalises on the inter-particle entanglements that occur out of locality. A shared key is generated through the execution of measurements on pairs of entangled particles that are distributed between the sender and receiver in this protocol. Compared to alternative protocols, E91 offers increased key rates and heightened security. The protocol utilises the principle of Bell's inequality and assesses secure key distribution by identifying violations of this inequality. By performing calculations on entangled particles, Alice and Bob can produce a sequence of stochastic digits that serve as the foundation of their secret key. Any attempt to intercept or quantify these particles generates a disturbance, thereby alerting Alice and Bob of the possible intrusion. Additionally, quantum error correction techniques are incorporated into the E91 protocol to ensure the accuracy and dependability of the transmitted data. These techniques assist in reducing the impact of noise and decoherence that may occur during quantum communication.

The E91 protocol operates on the fundamental tenets of quantum entanglement to establish a secure channel of communication between two entities. In the following, the fundamental stages encompassing the protocol are outlined.

1. Alice produces a pair of entangled particles, typically photons, and transfers one to Bob while retaining the other.
2. Alice and Bob proceed independently to select measurement parameters for their respective particles, including the polarisation basis.
3. Alice and Bob conduct comparisons and measurements on their particles following the parameters they specify. The measurement outcomes are documented, while the selected parameters remain undisclosed.
4. Alice and Bob perform calculations to ascertain whether Bell's inequality has been violated by comparing a subset of their measurement outcomes. A breach signifies that their particles have become entangled, thereby enabling the secure distribution of keys.
5. Alice and Bob conduct additional analysis and processing on the results of their measurements to derive a secure key suitable for encryption.

By adhering to these procedures, Alice and Bob can establish a mutually recognised secret key via the E91 Protocol, thereby guaranteeing secure communication that is resistant to interception and eavesdropping. Attempts to measure or intercept the entangled particles will disrupt their state, thereby alerting the communicating parties that an eavesdropper is present. The E91 protocol provides unconditional security, which means that regardless of the eavesdropper's computational capacity or available resources, the security of the communication channel is guaranteed. Adequate security measures are beyond the capabilities of traditional cryptographic techniques. Another advantage of the E91 protocol is its ability to support communication over extended distances. Despite the physical separation of the entangled particles over long distances, their entangled properties continue to exhibit correlation.

The implementation of the E91 protocol faces technical challenges that require advanced technology and specialised knowledge. The generation, handling, and measurement of entangled particles present practical challenges due to the need for thorough experimental configurations and precise control. In comparison to conventional methods, the E91 protocol exhibits comparatively slow communication rates. Delay in transmissions can be attributed to the probabilistic nature of quantum measurements and the necessity for error correction. Noise and decoherence are examples of environmental disturbances that quantum systems are susceptible to. These disturbances have the potential to cause errors and affect the integrity of the information being transmitted.

The potential applications of the E91 protocol in the domains of quantum communication and cryptography are promising. Some significant use cases include:

1. Secure communication: Even in the presence of potential eavesdroppers, the protocol enables the establishment of secure communication channels between two communicating parties. This capability enhances its applicability as a

method for conducting secure financial transactions, confidential data transmission, and messaging.

2. QKD framework: The E91 protocol functions can serve as a QKD system. By enabling the secure exchange of cryptographic keys, E91 provide a method to protect the confidentiality of sensitive data and preserve its integrity.

3. Quantum networking: The E91 protocol can be deployed in quantum networks to enable secure communication among network entities separated by large distances.

4. Quantum computing: In a field where secure communication and encryption are crucial, the E91 protocol can aid in the advancement of secure quantum protocols and algorithms that are impervious to quantum computer intrusions.

By implementing an unprecedented level of security, the E91 protocol holds the potential to fundamentally transform numerous industries. For example, secure communication is of the utmost importance in the healthcare industry as it facilitates the transfer of patient information, medical records, and telemedicine application data. An additional layer of security can be provided by the E91 Protocol, which safeguards patient confidentiality and prevents unauthorised access to sensitive medical data. Banking and finance are another application domain for E91 to help the finance sector meet standards to secure communication channels to facilitate transactions. By implementing the E91 protocol, the confidentiality of customer information can be maintained and unauthorised access to financial transactions prevented. Similarly, military and government organisations are entrusted with highly classified information that demands the highest level of security. Establishing secure communication channels for the transmission of classified data and encrypted communication between military personnel and government officials is possible via the implementation of the E91 Protocol. With the growing dependence on cloud computing and data centres, implementing the E91 Protocol, data transmission security can be improved, guaranteeing the preservation of confidentiality and integrity for sensitive information stored in the cloud.

Several misconceptions about the E91 protocol are being circulated in the literature. For example, instantaneous communication is not possible via the E91 protocol due to constraints imposed by the speed of light, i.e., information cannot be transmitted faster than the speed of light by the entangled particles, and distance does not affect their correlation. Additionally, although the E91 protocol provides absolute security, it is not immune to every conceivable attack. Security vulnerabilities may exist due to implementation deficiencies, limitations of technology, or side-channel attacks. Continuous research and development are vital to guarantee the resilience of the protocol against potential threats. The E91 protocol should not be mixed up with the concept of quantum supremacy, which pertains to the capacity of quantum computers to surpass classical computers in particular computations. The E91 protocol prioritises secure communication over processing speed.

Advancements in the domain of quantum communication have prompted research into the adoption and development of the functionalities of the E91 protocol. Possible future revisions and developments include the following:

1. Current research efforts are currently focused on optimising the E91 protocol's efficiency with the aim of augmenting communication rates while reducing resource requirements. This demands developments in measurement technologies, error correction methods, and entangled particle generation.
2. Potential future advancements could consider the incorporation of quantum repeaters into the E91 protocol. Quantum repeaters are specifically engineered devices to expand the range of secure quantum communication. By overcoming the constraints imposed by distance and decoherence, they facilitate secure communication across greater distances.
3. Interoperability and standardisation are imperative considering the progress being made in quantum communication technologies. There are ongoing efforts to establish universally accepted protocols, hardware interfaces, and measurement standards to guarantee the compatibility and smooth integration of quantum communication systems.
4. The increasing prevalence of quantum computers raised concerns about the security strength of traditional cryptographic techniques. Potential future advancements could involve the exploration of how the E91 protocol could be integrated with post-quantum cryptographic algorithms in order to ensure secure communication that is immune to quantum computer attacks.

6.3.3 DECOY-STATE QKD

Decoy-State QKD [13] is an innovative quantum communication mechanism with the potential to fundamentally transform the domain of secure information exchange. The protocols introduce the utilisation of decoy photons during information transmission, acting as a security mechanism to identify possible eavesdroppers. Decoy photons, which are generated at random, are transmitted along with the quantum states that contain the cryptographic keys. Eavesdroppers can be identified through the comparison of the statistics of the decoy photons received with the anticipated values. This mechanism offers many benefits in comparison to conventional encryption approaches, such as enhanced rates of key generation and higher resilience against a wide range of attack methods.

Decoy-State QKD works following the key tenets of quantum mechanics, making particular use of the characteristics of quantum states, including superposition and entanglement. The mechanism starts with the production of quantum states, which are commonly represented as particles and contain the cryptographic keys. These photons are subsequently transmitted to the receiving entity via a quantum channel. To improve security, Decoy-State QKD employs decoy photons. Decoy photons are deliberately designed to have a lower intensity in comparison to the signal photons, thus increasing their susceptibility to interception. Randomly generated decoy photons are transmitted along with the actual quantum states.

On the receiving side, the receiver measures and records the properties of the incoming particles. By comparing the statistical characteristics of the photons that have been received, encompassing both the signal and decoy photons, with the anticipated values, any anomalies resulting from interception attempts can be discovered.

By implementing this detection mechanism, an extra level of security is introduced, ensuring that the transmitted cryptographic keys maintain their integrity.

This mechanism offers many benefits in comparison to conventional encryption approaches, such as enhanced rates of key generation and higher resilience against a wide range of attack methods. A significant advantage of Decoy-State QKD is its enhanced key generation rates. The utilisation of decoy photons in the mechanism enables it to optimise the process of key generation, leading to increased key rates in comparison to conventional QKD protocols. Additionally, Decoy-State QKD exhibits enhanced resilience against a multitude of attack methods, such as photon number splitting. By employing decoy photons of varying intensities, an attempt to intercept the transmitted quantum states can be detected. The capacity to identify eavesdropping attempts serves to strengthen the communication channel's overall security.

Decoy-State QKD advantages made them attractive in applications that prioritise secure communication. Integrating it into a secure communication system necessitates a detailed evaluation of multiple parameters. Having a dependable source of quantum states is crucial for creating cryptographic keys. This source should have the ability to generate high-fidelity quantum states with minimum noise and mistakes. A secure quantum channel is necessary to transmit quantum states between the two parties. The channel must be protected from external interference and reduce loss and noise to maintain the quality of sent signals. A strong detecting system is required at the receiving end to measure and analyse the incoming photons. This system must differentiate between signal photons and decoy photons to detect potential eavesdroppers. Furthermore, error correction and privacy amplification methods must be used to guarantee the security and secrecy of the transmitted cryptographic keys. One such application area that exemplifies the secure communication need is telecommunications, where ensuring secure data transmission is of ultimate importance. Decoy-State QKD offers a strong method for creating secure communication channels and protecting sensitive data from interception and hackers. Similarly, the finance sector can derive significant advantages from Decoy-State QKD. The protocol's improved security features can protect financial data and prevent unauthorised access to sensitive information in light of the growing dependence on digital transactions and the importance of secure financial communication. Decoy-State QKD also has the potential to transform how government sectors and defence agencies create secure communication networks for exchanging sensitive information. The protocol's capability to identify eavesdroppers and withstand different hacking methods makes it a crucial tool for safeguarding confidential information and sustaining national security.

Despite its several benefits, Decoy-State QKD faces obstacles and restrictions. One major challenge is the need for rigorously defined experimental setups. Decoy-State QKD depends on accurate control and measurement of quantum states, which can be difficult to achieve in practical situations. Ensuring precise quantum operations and reducing noise and errors are essential for the success of the protocol. Decoy-State QKD is characterised by its vulnerability to channel loss and noise. Quantum states are susceptible to environmental factors including transmission loss and background noise, which can lower the quality of received signals. Extensive error correction and privacy amplification measures are necessary to reduce these impacts, increasing the complexity of the entire design. Additionally, Decoy-State QKD is currently constrained by its communication distance range. Photons undergo loss and deterioration as they traverse a quantum channel. Over extended distances, these effects can greatly affect the quality of the transmitted quantum states, resulting in a reduction in key generation rates. Addressing these constraints is a topic of continuous investigation in the field.

Decoy-State QKD is a dynamic area of study, where continuous research is focused on enhancing its functionalities and overcoming its constraints. Recent improvements have focused on improving detection techniques to strengthen the protocol's security and performance. Today, researchers are exploring the combination of Decoy-State QKD with other quantum technologies, like quantum repeaters, to enhance the scope of secure communication across extensive distances. By reducing the impact of channel loss and degradation, these developments could facilitate the broad use of Decoy-State QKD in many applications. Additionally, efforts are underway to simplify the implementation and enhance the scalability of Decoy-State QKD systems. The protocol can be made more accessible to various sectors and applications by simplifying its complexity and reducing resource requirements.

6.3.4 MEASUREMENT-DEVICE-INDEPENDENT QKD

Measurement-Device-Independent Quantum Key Distribution (MDI-QKD) protocol is a unique technique in quantum cryptography. MDI-QKD proposed by Lo, Curty, and Qi in 2012 [14], offers higher security guarantees by removing the need for trusted measurement devices. This method enables two communicating entities to establish a securely encrypted key even when the communication channel is not secure. In MDI-QKD, the communicating parties perform measurements independently, and the security of the protocol is based only on the laws of quantum mechanics. It achieves this by using entangled states and measurements performed at an untrusted node. This approach removes vulnerabilities associated with compromised or faulty measurement devices. It efficiently mitigates detector vulnerabilities and allows for a doubling of the safe transmission distance compared to conventional technologies.

The fact that MDI-QKD can operate effectively even in situations with significant signal loss and using commonly available optical components makes it a robust protocol for quantum cryptography systems. Initially introduced in 2011, this protocol provides substantially greater rates of generating encryption keys compared to fully device-independent QKD models. MDI-QKD promises an efficient approach for long-distance quantum cryptography without requiring near-perfect detection efficiency or advanced qubit amplification.

MDI-QKD guarantees that two parties can create and exchange encryption keys securely. Below is a simplified explanation of the protocol:

6.3.4.1 Key Generation

Alice and Bob transmit quantum states to Charlie, a neutral third party. Charlie performs measurements on these quantum states and publicly declares the outcomes. Using this outcome, Alice and Bob then generate encryption keys that are resistant to interception or manipulation by measurement devices. The key generation process involves the following steps:

1. Entanglement Generation: The process starts by creating entangled quantum states between Alice and Bob. These interconnected states act as the foundation for secure key distribution. Alice and Bob individually produce one half of a pair of entangled particles, such as photons, using a quantum source.
2. State Preparation and Transmission: Once Alice and Bob have created their own entangled pairs, they proceed to prepare quantum states using various polarisation or phase encoding techniques. Subsequently, these states are sent to the "quantum referee" Charlie, who performs joint measurements on the received states
3. Measurement Phase: Charlie obtains the quantum states from both Alice and Bob and makes joint measurements on them. The joint measurements are essential for establishing correlations between the states transmitted by Alice and Bob. Charlie's measurements depend on the encoding techniques employed by Alice and Bob.
4. Bell Test: During the measurement phase, Charlie can run a Bell test to confirm the existence of entanglement between the quantum states transmitted by Alice and Bob. Deviation from the anticipated outcomes in the Bell test can suggest possible eavesdropping attempts.
5. Information Exchange: Following the measurement phase, Alice and Bob exchange classical information over a public channel. This information is typically in the form of measurement outcomes or selections of measurement bases. This data is used to choose a subset of correlated measurement outcomes, confirming the presence of a secure key.
6. Key Distillation: Alice and Bob then use traditional error correction and privacy amplification techniques on the correlated measurement outcomes to extract a secure cryptographic key. These techniques guarantee that any information leaked to Eve during the key distribution process is insignificant.
7. Authentication: Alice and Bob authenticate the quantum communication channel and confirm each other's identities to guarantee that the key exchange is protected and restricted to the intended participants.
8. Key Establishment: After the key extraction process and successful authentication, Alice and Bob have established a mutually agreed secret key that may be deployed for secure communication.

6.3.4.2 Security and Device Independence

The protocol is specifically designed to be compatible with any device, ensuring protection against any possible weaknesses in the measuring hardware. This is accomplished by focusing on the correlations between signals originating from Alice and Bob, rather than the signals themselves. Alice and Bob adjust the signal polarisation or phase to encode bit values, which are subsequently assessed by Charlie in terms of interference rather than the qubits themselves.

6.3.4.3 Advancements in Technology and Security

MDI-QKD incorporates innovative protocols such as the symmetric three-intensity and four-intensity decoy-state schemes to effectively eliminate any potential information leaks in the measurement unit, providing a strong defence against quantum attacks. The integration of machine learning to address real-time phase compensation and the implementation of mode-pairing to overcome global phase locking complications demonstrate the innovative adaptations in MDI-QKD systems.

When comparing MDI-QKD with other QKD protocols, many significant differences arise that emphasise the advancements and challenges facing quantum cryptography. These are summarised as follows:

- System complexity and efficiency: Alternative QKD protocols usually need software that scans and transmits data or the addition of an extra device to calibrate the phase reference frame. This leads to greater complexity and a decrease in transmission efficiency. On the other hand, asynchronous MDI-QKD, also known as mode-pairing MDI-QKD, achieves a comparable repeater-like rate-loss scaling but with the added benefit of simpler technological implementation, making it a more efficient alternative option.
- Key rate and real-time encryption: Experimental results demonstrated that asynchronous MDI-QKD achieves the maximum key rate while maintaining MDI security over distances ranging from 50 km to 480 km. At fibre distances of 50 km and 100 km, the key rates achieve 6.02 Mbps and 2.29 Mbps respectively, allowing for real-time video encryption using a one-time-pad [15].
- Security assurances and technology advancements: Although alternative QKD has advantages in terms of simplicity, clock rates, and detection efficiencies, MDI-QKD offers a more robust security assurance, particularly against attacks targeting the detectors. The emergence of 'twin-field' QKD presents the potential to extend communication distance while preserving the same level of security as MDI-QKD, without introducing considerable complexity to the setup.
- MDI-QKD, similar to other QKD protocols, encounters challenges such as limited transmission distance, implementation complexity, and the need for seamless integration with current infrastructure. MDI-QKD suffers from several challenges and limitations, including:
- Challenges in aligning polarisation: Polarisation alignment is a substantial obstacle in MDI-QKD, since any deviation in the state of polarisation can greatly impact the efficiency of key generation. Additionally, traditional approaches to dealing with polarisation alignment require extra resources

or may interrupt the QKD process, resulting in a reduction in system efficiency. A new method for aligning polarisation in MDI-QKD exploits inherent events from MDI to directly compute and rectify polarisation deviation. This approach incorporates the polarisation dimension with the encoding dimension, and develops a response-rate model for MDI-QKD that includes additional polarisation dimensions [16].

- Environmental factors that impact the distance of transmission: MDI-QKD may achieve a maximum secure key transmission distance of 178 kilometres when the conditions are clear, by utilising vector vortex beams. Yet, environmental conditions such as rainfall and fog reduce this distance.

- Experimental Complexity: Due to the need for precise control over quantum states, entanglement generation, and joint measurements, experimental implementation of MDI-QKD can be extremely complicated. Complex experimental configurations and technologies are often needed to meet these requirements, which is costly and difficult to maintain.

- Efficiency and Rate: In comparison to alternative QKD protocols, MDI-QKD generally runs at a reduced key generation rate. This is partially due to the requirement for additional cycles of joint measurements and communication involving a third party (Charlie). The practical feasibility of MDI-QKD in situations that demand high-throughput secure communication might be constrained by the use of lower key rates.

- Synchronisation Conditions: The MDI-QKD protocol necessitates accurate synchronisation between Charlie, Alice and Bob. It can be difficult to achieve and sustain this synchronisation over long distances and in the presence of network fluctuations or delays.

- Detection Efficiency and Loss Tolerance: The performance of MDI-QKD is significantly influenced by the efficiency of the detectors employed for quantum state measurement. A low detection efficiency or high loss in the communication channel can have a substantial effect on the key generation rate and security of the protocol. To tackle these constraints, it is often required to employ complex detection technologies and mitigation strategies.

- Vulnerability to Side-Channel Attacks: Despite MDI-QKD capability to protect against specific types of threats, it remains susceptible to side-channel attacks and other methods that exploit implementation vulnerabilities. To ensure the overall security of MDI-QKD, potential attack vectors must be thoroughly evaluated and suitable countermeasures must be implemented.

- Practical Implementation Challenges: The deployment of MDI-QKD in real-world scenarios presents practical implementation challenges, including but not limited to compatibility with standard protocols, integration with pre-existing communication infrastructure, and scalability to accommodate large-scale networks. To tackle these challenges, scholars specialising in network engineering, cryptography, and quantum physics must engage in interdisciplinary collaboration.

- Resource Requirements: The MDI-QKD algorithm needs substantial computational resources to execute error correction, privacy amplification, and key distillation, particularly in situations involving high channel noise or loss. For practical deployment, it is critical to achieve a balance between the required resources and the desired level of security.
- Notwithstanding these obstacles, ongoing advancements in technology and research continue to rectify numerous constraints associated with MDI-QKD, thus positioning it as a promising protocol to provide secure communication within quantum networks. MDI-QKD may, nevertheless, demand additional progress in experimental methodologies, technological innovation, and standardisation efforts to be practically implemented and widely appreciated.

In 2022, the viability of distributing quantum-encrypted keys via MDI-QKD protocols under real-world conditions was demonstrated by two separate research groups. The first group of scientists from academic institutions in the United Kingdom, France, and Switzerland, accomplished a significant milestone by generating 1.5 million entangled Bell pairs between a pair of trapped strontium-88 ions that were separated by a distance of two meters [17]. This experiment produced a shared key with a length of 95,884 bits, thereby demonstrating the practical capabilities of MDI-QKD to secure communications. The second group, of researchers from the National University of Singapore and Ludwig-Maximilian University in Germany, utilised optically trapped rubidium-87 atoms that were stored in laboratories separated by 400 meters [18]. A 700-meter-long optical fibre was utilised to maintain an entanglement fidelity of 89.2% and a quantum bit error rate of 7.8%. This experiment provides additional evidence of the resilience and viability of MDI-QKD in facilitating the transmission of quantum-encrypted keys across large distances.

6.3.5 CONTINUOUS-VARIABLE QKD

Continuous-Variable Quantum Key Distribution (CV-QKD) [19] enables secure communications using the continuous properties of quantum states, such as the amplitude and phase of light. CV-QKD uses light beams carrying an infinite number of quantum states, as opposed to discrete-variable QKD protocols that encode information onto individual quantum states. The utilisation of this quantum advantage strengthens the protocol's security and efficacy.

In CV-QKD, the continuous variables pertain to characteristics including the amplitude and phase of light. CV-QKD enables the transmission of quantum states to be utilised to establish a secure key between two parties. By using this key to encrypt and decrypt messages, access to the information is restricted to authorised parties only.

CV-QKD operates on the fundamental tenets of quantum mechanics, which dictate the quantum-level behaviour of particles. Particles can persist in superposition, in which they hold multiple states concurrently, as described by quantum mechanics. This property enables the encoding of information in quantum states, providing an eavesdropping-resistant and secure method of communication.

The CV-QKD protocol comprises three main elements, namely, the sender, the receiver, and the quantum channel. The transmitter encodes and transmits the information in quantum states over the quantum channel. The receiver decodes the received quantum states to recover the original information. The quantum channel is the medium for the transmission of quantum states. CV-QKD requires standard channels of communication between the sender and receiver. Information required for the key generation process, including parameter estimation and error correction, is transmitted through these channels.

The operation of the CV-QKD protocol can be divided into five phases: initialisation, estimation of parameters, generation of keys, error correction, and amplification of privacy.

- During the initialisation phase, the quantum channel is prepared and a common reference frame between the sender and receiver is established. This provides accurate encoding and decoding of quantum states.
- Once the protocol is initialised, parameter estimation proceeds to define the quantum channel's properties. This enables the sender and the receiver to identify the most effective settings for encoding and decoding quantum states.
- Next, the key generation phase begins. The sender encodes information into quantum states and transmits it to the receiver via the quantum channel. The key is then extracted by the receiver through quantum state measurements.
- Error correction is executed after key generation to correct any errors that might have occurred during the transmission. The process entails the comparison of the keys obtained by the sender and receiver, followed by the application of suitable error correction codes to correct any inconsistencies.
- Finally, privacy amplification is executed to strengthen the key's security. Privacy amplification involves distilling a shortened yet more secure key from the original key by executing operations that eliminate any information that might have been leaked during transmission.

CV-QKD offers many advantages in terms of simplicity and compatibility with existing communication infrastructure, making it a promising candidate for practical implementations. Its use of continuous variables, which enables the transmission of an infinite number of quantum states, is one of its primary advantages. This offers a greater level of security compared to discrete-variable QKD protocols, which depend on particles as components. CV-QKD's compatibility with existing fibre optic infrastructure is an additional benefit. The use of standard optical components in the implementation of CV-QKD facilitates its seamless integration into pre-existing communication networks. The cost-effectiveness and practicality of CV-QKD as a secure communication solution are derived from its compatibility. Additionally, CV-QKD provides a faster information transmission rate than alternative QKD protocols. By facilitating the transmission of larger quantities of data per quantum state, CV-QKD achieves a faster and more efficient communication method.

Despite its many advantages, CV-QKD has some limitations and challenges. Noise in the quantum channel is among the primary obstacles. During transmission, noise may introduce errors that compromise the security and accuracy of the key generation process. Ongoing research efforts are focused on designing techniques that reduce the impacts of noise and improve the overall performance of CV-QKD. An additional constraint associated with CV-QKD is its requirement for a trusted quantum channel. CV-QKD's security is built on the assumption that an eavesdropper cannot compromise or tamper with the quantum channel. Hence, it is imperative to ensure the integrity of the quantum channel. CV-QKD also needs accurate synchronisation and calibration between the sender and receiver. Inconsistencies in the settings or timing may result in errors and affect the protocol's performance. Achieving precise calibration and synchronisation is challenging, particularly when applied in real-world scenarios.

6.4 DISCUSSION AND FUTURE PROSPECTS

Each of these QKD protocols offers unique advantages and limitations concerning efficiency, practicality, and security. A comprehensive understanding of the characteristics and basic principles of these protocols is imperative to build secure communication systems and assess their suitability for particular applications.

The BB84 protocol is a frequently used QKD system that leverages quantum state features to safely communicate cryptographic keys. The BB84 protocol does not utilise decoy photons and depends on exchanging quantum states across a quantum channel. BB84 is easier to set up but may be more vulnerable to specific hacking techniques than Decoy-State QKD. The entanglement-based QKD, E91 protocol, depends on creating and measuring entangled photon pairs to form secure communication pathways. The E91 protocol has benefits such as enabling long-distance quantum communication, but it is more intricate to execute than Decoy-State QKD. Table 6.1 presents a summary of the essential attributes and limitations of the reviewed QKD protocols.

Looking forward, QKD protocols are set to find applications in various sectors, including secure communication networks, quantum Internet, banking and finance, healthcare, and government sectors. As quantum technologies continue to mature, the integration of QKD into existing communication infrastructure will become more viable, paving the way for a new era of secure and resilient communication systems.

While challenges persist, the outlook of QKD is promising, with significant potential to transform secure communication and mitigate cybersecurity concerns in the quantum era. With ongoing technological advancements and research, QKD protocols are positioned to significantly influence the trajectory of secure communication in the future.

6.5 CONCLUSION

In conclusion, Sections 6.1, 6.2, and 6.3 provided a comprehensive overview of QKD, describing its fundamental principles, historical context, and various protocols. Section 6.1 explored the fundamental principles of quantum mechanics that underpin QKD, focusing on the unique features of quantum states and their relevance to

Table 6.1
Comparison of the Studied QKD Protocols

Protocol	Overview	Key Limitations
BB84 Protocol	Relies on the transmission of photons encoded with quantum states, such as polarisation, and subsequent measurement.	Vulnerable to photon loss and channel noise, limiting transmission distances.
E91 Protocol	Utilises the phenomenon of quantum entanglement to establish secure communication channels.	- Requires a reliable source of entangled particles, which is challenging to implement. - Vulnerable to noise and decoherence, affecting the fidelity of entanglement.
Decoy-State QKD	Enhances security and efficiency by incorporating additional decoy states into the key generation process.	- Increased computational complexity. - Sensitivity to errors in decoy state preparation and detection.
MDI-QKD	Offers enhanced security by removing the need for trusted measurement devices.	- Requires advanced quantum setups and synchronisation techniques. - Relatively low key generation rates.
CV-QKD	Utilises the continuous properties of quantum states, such as the amplitude and phase of light.	- Susceptible to Gaussian noise and loss, limiting transmission distances. - Requires precise calibration and synchronisation of measurement devices.

secure communication. Section 6.2 examined the historical evolution of QKD, detailing its historic progression from theoretical concepts to practical implementations, and illustrating its journey from a theoretical concept to a promising technology. In Section 6.3, the prospects of QKD protocols and their applications were discussed.

Despite the challenges and limitations faced by current QKD protocols, such as susceptibility to noise, limited transmission distances, and complex infrastructure requirements, ongoing research efforts aim to overcome these challenges and unlock the full capabilities of quantum communication. The protocols discussed in Section 6.3 present diverse approaches to secure key exchange, each with its own strengths and weaknesses. The development of new protocols and advancements in existing ones has the potential to improve the security, efficiency, and practicality of QKD.

REFERENCES

1. Stephen Wiesner. Conjugate coding. *SIGACT News*, 15(1):78–88, 1983.

2. *Quantum cryptography: Public key distribution and coin tossing*, volume 175, 1984.

3. Albert Einstein, Boris Podolsky, and Nathan Rosen. Can quantum-mechanical description of physical reality be considered complete? *Physical Review*, 47(10):777, 1935.

4. Daniel M. Greenberger, Michael A. Horne, Abner Shimony, and Anton Zeilinger. Bell's theorem without inequalities. *American Journal of Physics*, 58(12):1131–1143, 1990.

5. Hugh Everett and JA Wheeler. Dewitt, bryce; graham, r. neill (eds.). the many-worlds interpretation of quantum mechanics. *Princeton Series in Physics*. Princeton, NJ: Princeton University Press. *pv ISBN 0-691-08131-X*, 1973.

6. Hugh Everett. *Hugh Everett theory of the universal wavefunction*. PhD thesis, Thesis, Princeton University, 1957.

7. Bryce S DeWitt. Quantum mechanics and reality. *Physics Today*, 23(9):30–35, 1970.

8. Wojciech Zurek and William Wootters. The no-cloning theorem. Technical report, Los Alamos National Lab. (LANL), Los Alamos, NM, 2008.

9. Werner Heisenberg. The actual content of quantum theoretical kinematics and mechanics. Technical report, 1983.

10. Werner Heisenberg. *Encounters with Einstein: And other essays on people, places, and particles*, volume 4, Princeton, NJ: Princeton University Press, 1989.

11. Lisa M Dolling, Arthur F Gianelli, and Glenn N Statile. *The tests of time: Readings in the development of physical theory*. Princeton, NJ: Princeton University Press, 2003.

12. Artur K Ekert. Quantum cryptography based on bell's theorem. *Physical Review Letters*, 67(6):661, 1991.

13. Hoi-Kwong Lo, Xiongfeng Ma, and Kai Chen. Decoy state quantum key distribution. *Physical Review Letters*, 94(23):230504, 2005.

14. Hoi-Kwong Lo, Marcos Curty, and Bing Qi. Measurement-device-independent quantum key distribution. *Physical Review Letters*, 108(13):130503, 2012.

15. Yuan-Mei Xie, Jun-Lin Bai, Yu-Shuo Lu, Chen-Xun Weng, Hua-Lei Yin, and Zeng-Bing Chen. Advantages of asynchronous measurement-device-independent quantum key distribution in intercity networks. *Physical Review Applied*, 19(5):054070, 2023.

16. Jia-Xuan Li, Guan-Jie Fan-Yuan, Shuang Wang, Ze-Hao Wang, Feng-Yu Lu, Zhen-Qiang Yin, Wei Chen, De-Yong He, Guang-Can Guo, and Zheng-Fu Han. Polarization alignment in measurement-device-independent quantum key distribution with intrinsic events. *Physical Review Applied*, 20(5):054041, 2023.

17. David P Nadlinger, Peter Drmota, Bethan C Nichol, Gabriel Araneda, Dougal Main, Raghavendra Srinivas, David M Lucas, Christopher J Ballance, Kirill Ivanov, EY-Z Tan, et al. Experimental quantum key distribution certified by Bell's theorem. *Nature*, 607(7920):682–686, 2022.

18. Wei Zhang, Tim van Leent, Kai Redeker, Robert Garthoff, René Schwonnek, Florian Fertig, Sebastian Eppelt, Wenjamin Rosenfeld, Valerio Scarani, Charles C-W Lim, et al. A device-independent quantum key distribution system for distant users. *Nature*, 607(7920):687–691, 2022.

19. Lars S Madsen, Vladyslav C Usenko, Mikael Lassen, Radim Filip, and Ulrik L Andersen. Continuous variable quantum key distribution with modulated entangled states. *Nature Communications*, 3(1):1083, 2012.

7 Quantum Key Distribution, Security, and Analysis

Mohammad Hammoudeh, Clinton M. Firth,
Ahcene Bounceur, Bamidele Adebisi,
and Abdullah T. Alessa

7.1 INTRODUCTION

Anticipated future applications for QKD protocols include a wide range of fields, such as government operations, banking and finance, secure communication networks, and quantum internet. The integration of QKD into current communication infrastructure will become more viable as quantum technologies advance, resulting in a paradigm shift towards communication systems that are both secure and resilient.

The discussions around classical versus quantum security methodologies gained considerable attention in the field of cybersecurity since the advent of quantum computing technologies. Classical security methods, which depend on computational complexity and mathematical algorithms to ensure the confidentiality, integrity, and authenticity of data, have long been the foundation of modern cryptography. However, the emergence of quantum computing presents a substantial threat to traditional cryptographic systems. This is because quantum algorithms, including Shor's algorithm, can efficiently factor large numbers and break prevalent encryption schemes such as Rivest–Shamir–Adleman (RSA) and Elliptic-Curve Cryptography (ECC). On the other hand, quantum security approaches, which are founded on the principles of quantum mechanics, hold the potential to provide protection and are theoretically unbreakable, even against quantum adversaries.

Classical security techniques, such as symmetric and asymmetric encryption, digital signatures, and hash functions, were established as the basis of secure communication. To ensure security, these methods rely on mathematical principles including discrete logarithms, prime factorization, and elliptic curve mathematics. Although classical cryptographic systems have demonstrated efficacy against adversaries employing traditional attack techniques, their susceptibility to quantum computer attacks stems from the intrinsic computational constraints of classical algorithms. The advancement of quantum computing technology raises concerns about the security of conventional cryptographic systems, thus prompting the investigation of alternative security techniques.

DOI: 10.1201/9781003475286-7

Quantum security techniques, QKD protocols, implement information-theoretic security guarantees based on the principles of quantum mechanics. In contrast to conventional cryptographic systems that depend on computational complexity and mathematical assumptions, quantum security protocols provide security assurances based on the principles of physics. QKD protocols specifically leverage features including the uncertainty principle, no-cloning theorem, and quantum entanglement to enable secure communication. Due to their adoption of the fundamental properties of quantum mechanics, QKD protocols are immune to both classical and quantum adversary attacks.

When comparing classical and quantum security approaches, it is essential to consider that information-theoretic security guarantees are provided by quantum security protocols, as opposed to classical cryptographic systems which rely on the complexity of mathematical problems for security. Additionally, conventional cryptographic systems are susceptible to quantum computer attacks, of which large numbers can be efficiently factorized, leading to a breach of prevalent encryption schemes. On the contrary, quantum security protocols provide an unbreakable level of protection against intrusions initiated entirely through computational methods, protecting against attacks from both classical and quantum adversaries. It is also important to consider the practical applications of classical and quantum security approaches. Conventional cryptography is a mature and well-established security infrastructure that has been the subject of extensive research and widespread deployment. Although still in their infancy, quantum security protocols show potential for enabling secure communication within upcoming quantum networks. However, the implementation of quantum security protocols in practical settings faces challenges including constrained transmission ranges, infrastructure demands, and susceptibility to specific forms of attacks. Table 7.1 provides a comparison between classical and quantum security approaches, showing their respective strengths, vulnerabilities, practical implementations, and prospects.

Table 7.1

Comparative Analysis between Classical and Quantum Security Approaches

Aspect	Classical Security	Quantum Security
Security guarantees	Based on mathematical complexity	Information-theoretic, rooted in quantum mechanics
Vulnerabilities	Vulnerable to attacks from quantum computers	Immune to attacks from both classical and quantum adversaries
Practical implementations	Mature and well-established infrastructure	Still in early stages of development, facing challenges such as limited transmission distances and infrastructure requirements
Future prospects	Threatened by advancements in quantum computing	Holds promise for secure communication in future quantum networks, ensuring confidentiality, integrity, and authenticity of information

The ongoing advancement of quantum computing technology will bring about changes in the security setting, which will require the implementation of quantum security methodologies to mitigate emerging risks. Quantum security protocols ensure the confidentiality, integrity, and authenticity of data in the face of ever-evolving threats, hence promising secure communication in the quantum era. The discussion on classical versus quantum security methodologies emphasizes the imperative for innovative resolutions to address emerging cybersecurity challenges. For decades, classical cryptographic systems have been integral to ensuring secure communication. However, with the emergence of quantum computing, this foundation has been significantly weakened. Quantum security approaches, founded upon the principles of quantum mechanics, offer the potential for unmatched security guarantees immune to emerging attacks from adversaries of both classical and quantum nature. The continued development of quantum technologies demands the implementation of quantum security protocols to safeguard the confidentiality and integrity of communications in the quantum age.

This chapter provides a comprehensive investigation of QKD's role in enabling secure communication. It provides security analysis of common QKD protocols, analysing the principles that underpin its robustness against eavesdropping and other security threats. Then, it addresses the operational principles and implementation challenges of QKD protocols, highlighting the advancements made in practical deployments and the ongoing efforts to overcome technical limitations. This chapter concludes by examining the applications and prospects of QKD, showcasing its potential to revolutionize secure communication networks, quantum internet, banking and finance, healthcare, and government sectors.

7.2 QUANTUM-RESISTANT ALGORITHMS AND PROTOCOLS

A crucial principle of quantum security is the use of quantum-resistant algorithms. Quantum-resistant algorithms, post-quantum, or quantum-safe cryptographic algorithms are specifically engineered to endure attacks originating from classical and quantum computers. Conventional cryptographic algorithms, such as RSA and ECC, are vulnerable to attacks by quantum algorithms, such as Shor's algorithm, which solves discrete logarithm problems and efficiently factors large numbers. In contrast, quantum-resistant algorithms are explicitly engineered to withstand attacks originating from quantum computers, thus securing confidential data in the era of quantum computing.

Quantum-resistant algorithms can be developed using a variety of techniques, such as isogeny-based cryptography, code-based cryptography, hash-based cryptography, multivariate polynomial cryptography, and lattice-based cryptography. These algorithms apply mathematical problems that are considered to be computationally hard for both classical and quantum computers to efficiently solve. Lattice-based cryptography, for instance, depends on the difficulty of decoding linear codes and error-correcting codes, whereas code-based cryptography relies on the difficulty of decoding lattice problems, such as the shortest vector problem.

In addition to quantum-resistant algorithms, quantum security also relies on the development of quantum-resistant protocols, also referred to as post-quantum protocols. These cryptographic protocols are specifically designed to enable secure data exchange and communication even in the presence of quantum computing adversaries. To withstand quantum attacks, quantum-resistant protocols generally employ cryptographic algorithms that are considered to be secure. The imminent arrival of quantum computers with the ability to breach conventional cryptographic systems like RSA and ECC increased the demand for protocols that can resist assaults from both classical and quantum adversaries. The objective of quantum-resistant protocols is to guarantee the confidentiality, authenticity, and integrity of data exchange and communication. The primary aim of these protocols is to provide security guarantees that maintain their validity despite the threat from powerful quantum adversaries.

Ensuring the efficiency and practicality of quantum-resistant algorithms and protocols is a critical obstacle in their development and implementation. A considerable amount of computational resources is needed to implement several quantum-resistant algorithms and protocols, which may hinder their applicability in the real world. Furthermore, the process of migrating from classical to quantum-resistant algorithms may require a complex and time-intensive revision of current cryptographic infrastructure and protocols.

Regardless of these obstacles, quantum-resistant algorithms and protocols are of the utmost importance in protecting sensitive data against emerging quantum threats over the long term. The advancement of quantum computing technology calls for the development and implementation of quantum-resistant algorithms and protocols to ensure the confidentiality, integrity, and authenticity of data and communications in the quantum era.

7.3 QUANTUM-RESISTANT INFRASTRUCTURE

Quantum-resistant infrastructure refers to the hardware, software, and network components designed to enable secure communication and data exchange in the presence of quantum computing adversaries. With the potential advent of quantum computers capable of breaking traditional cryptographic systems (e.g. RSA and ECC), there is an increased demand for infrastructure that can withstand attacks from both classical and quantum adversaries.

Quantum-resistant infrastructure comprises various components, including quantum-resistant cryptographic algorithms, quantum-resistant communication protocols, quantum-resistant hardware, and quantum-resistant network architecture. The purpose of these components is to provide security guarantees in the presence of quantum adversaries.

Quantum-resistant cryptographic algorithms form the foundation of quantum resistance infrastructures. In addition to cryptographic algorithms, quantum-resistant infrastructure includes communication protocols that enable secure communication and data exchange in quantum communication networks. Quantum-resistant infrastructure also comprises hardware components, such as quantum-resistant processors and cryptographic accelerators, designed to support the execution of quantum-resistant cryptographic algorithms. These hardware components incorporate specialized security features to protect against physical attacks and tampering. Finally, quantum-resistant infrastructure includes network architecture and management systems designed to support the deployment and operation of quantum-resistant communication networks. These systems ensure the reliability, scalability, and manageability of quantum-resistant networks, enabling organizations to securely communicate and exchange data in the quantum era.

7.4 ATTACKS ON QKD

While QKD offers security guarantees, there remain several limitations and challenges to resolve. One challenge is the practical implementation of QKD systems, which frequently encounter issues related to channel losses, noise, and the rate of key generation. These issues can limit the performance and efficiency of QKD systems in real-world applications. An additional challenge is the vulnerability of QKD to side-channel attacks. Side-channel attacks exploit information leaked through physical or implementation-related components of the QKD system, such as power analysis or timing information. QKD protocols are also vulnerable to various eavesdropping attacks that attempt to intercept and learn information about the exchanged quantum states. These attacks can potentially jeopardize the security of QKD systems, showing the demand for comprehensive security measures that go beyond the purely theoretical considerations. This section aims to provide a review of the significant attacks on QKD.

7.4.1 EAVESDROPPING ATTACKS

Eavesdropping attacks in QKD can take different forms, with attackers employing various techniques to intercept and decode transmitted information. In practical scenarios, an eavesdropper can extract crucial information from multi-photon states when employing attenuated laser radiation as the quantum state source. To mitigate potential eavesdropping, it is recommended to deploy cryptographic protocols such as decoy states or SARG04.

One of the common eavesdropping attacks is the intercept-resend attack, also known as the photon-number-splitting (PNS) attack. In this attack, the adversary intercepts the quantum states transmitted between the communicating parties, measures their properties to learn information about the key, and then sends an identical photon to Bob to the receiver. By doing so, the eavesdropper can gain partial or complete knowledge of the exchanged key without being detected.

Another type of eavesdropping attack is the beam-splitting attack, where the eavesdropper intercepts and splits the incoming photons into multiple paths, keeping one path for measurement while allowing the other path to continue to the receiver. This allows the eavesdropper to gain information about the key without being detected. In this attack, the eavesdropper exploits the nonzero multi-photon emission probability of the sender's laser source and the lossy nature of a quantum channel.

The PNS attack [1] is a targeted type of attack. It is applied to realistic photon sources emitting weak coherent pulses which generate single photons only with a certain probability. With a small probability, these sources emit multi-photon pulses containing two or more photons sharing the same polarization. The strategy for the adversary is to intercept these pulses coming from the sender, take one photon of the multi-photon pulse and send the remaining photon(s) along to Bob. The eavesdropper waits until the legitimate communicating parties publicly compare their measurement bases and then measures the intercepted photon on the correct basis.

The time-shift attack is another eavesdropping method [2]. The attacker takes advantage of the efficiency mismatch of two single photon detectors in a QKD system [3]. In this attack, the eavesdropper intercepts and measures the quantum states, and then generates a new time-shifted signal based on the measurement result before forwarding them to the receiver. This introduces delays in the transmission of quantum states between the legitimate communicating parties. By delaying the transmission, the adversary can learn information about the key without being detected by the legitimate parties. When there is a complete detector efficiency mismatch, i.e. there is a time window where the detector for the bit "0" is active while the detector for the bit "1" is completely inactive and vice versa, the adversary can obtain full information on the key without introducing any error.

To detect and mitigate the effects of eavesdropping attacks, QKD protocols employ a range of security measures. These include the implementation of privacy amplification techniques, randomization of measurement bases, and monitoring of error rates.

One of the common methods used to detect eavesdropping attacks in QKD is by monitoring the Quantum Bit Error Rate (QBER). QBER is the rate at which errors occur during the transmission of qubits. The QBER should fall below a specified threshold in a secure QKD system, signifying that the transmission is reliable and immune to eavesdropping. By continuous monitoring of the QBER, the sender and receiver can detect changes that may suggest an eavesdropper. An increase in the QBER indicates that the qubits have been disturbed, possibly by a measurement attempt. This initiates further investigation to ensure the security of the shared encryption keys.

Quantum state tomography [4] is a highly effective technique to detect eavesdropping attacks in QKD. Quantum state reconstruction determines the quantum state of transmitted qubits by measuring a set of identical quantum state data. By comparing the reconstructed state with the anticipated state, any discrepancies caused by eavesdropping can be detected. Quantum state tomography enables an in-depth analysis of the integrity of the quantum channel, offering valuable information about the possibility of any eavesdroppers. Complex mathematical algorithms and statistical techniques accurately detect minor alterations in the quantum state that could potentially flag an attack.

Although there has been considerable advancement in the detection of eavesdropping attacks in QKD, there remain unresolved challenges and limitations that require attention. A significant obstacle facing practical QKD systems is the presence of noise and imperfections, which can impact the accuracy of eavesdropping detection methods. Moreover, the difficulty of applying complex detection methods and the requirement for specialized equipment present logistical challenges. The scientific community are developing more robust and effective detection methods that can mitigate these challenges and facilitate secure communication in practical scenarios.

7.5   SIDE-CHANNEL ATTACKS

Side-channel attacks on QKD protocols exploit implementation vulnerabilities or physical properties of the system to leak sensitive information about the exchanged keys instead of directly intercepting the transmitted quantum states. These attacks specifically aim to exploit unintended channels, such as timing information, power consumption, electromagnetic radiation, or acoustic emissions, to leak information about its internal processes through unintended side channels. These side channels, which are often unexpected results of the system's operation, can be analysed by attackers to derive sensitive information, including the cryptographic keys used in QKD.

Side-channel attacks on QKD can take various forms, each exploiting distinct physical properties of the system. One common type of side-channel attack is a timing analysis which involves measuring and monitoring the timing of quantum operations or classical communications within the QKD system. By detecting variations in timing, attackers can derive information about the key generation process, potentially inferring information about the exchanged keys. Another type of side-channel attack is timing-based attacks, where adversaries observe fluctuations in power consumption within QKD hardware to leak information about cryptographic operations. Electromagnetic radiation analysis targets emissions from QKD hardware during operation. By intercepting and analysing electromagnetic radiation patterns, adversaries can glean insights into the key generation process, compromising system security. Similarly, acoustic analysis focuses on measuring and analysing acoustic emissions produced by QKD hardware. Changes in acoustic patterns may indicate specific protocol activities or key generation steps, providing adversaries with details to compromise the system's security. Electromagnetic radiation attacks, another form of side-channel attack, involve observing the electromagnetic emissions of the QKD device. By analysing the emitted radiation, adversaries can deduce sensitive information about the exchanged cryptographic keys.

Side-channel attacks present a serious threat to the security of QKD protocols and must be mitigated through robust security countermeasures and continuous vigilance to ensure the security of quantum communication channels [5]. Several countermeasures exist to mitigate side-channel attacks on QKD. One cryptographic countermeasure is to introduce randomness in the key generation process to make it difficult for attackers to deduce the keys from side-channel information. Using techniques such as masking, blinding, noise, and randomization obscure sensitive information and prevent side-channel leakage. Another approach is to continuously monitor and analyse the side-channel leakage of the QKD system. Understanding the emerging threats and potential sources of leakage, system designers can adopt measures to reduce side-channel risks and improve the security of cryptographic keys. Adopting the principles of secure by design in hardware and software can reduce the success of side-channel attacks. Designers can implement physical isolation mechanisms to minimize the impact of external monitoring on QKD hardware. Finally, conducting comprehensive security assessments and testing to detect and mitigate vulnerabilities in QKD implementations can prevent side-channel attacks.

7.6   OTHER ATTACKS ON QKD

Denial of service attacks against QKD systems can occur when a dedicated fibre optic line or line of sight in free space, connecting the two nodes involved in the QKD process, is intentionally disrupted by physical means, such as cutting or blocking the line. This vulnerability highlights the need for the development of QKD networks, which would redirect communication through alternative links in the event of such interruptions.

Within the context of Trojan-horse attacks on QKD systems, an attacker exploits the system by sending bright light into the quantum channel and analyses the resulting back-reflections. Recent research demonstrated that the adversary can derive the receiver's secret basis choice with a probability exceeding 90%, thereby compromising the security of the system [6]. The Trojan horse attack operates without the need for direct physical access to the endpoints. Instead of attempting to intercept individual photons transmitted between the endpoints, the adversary sends a significant burst of light back towards the sender during the transmission of photons. Some of the adversary's light gets reflected by the sender's equipment, thereby exposing the state of the sender's basis, e.g. as a polarizer. This form of attack can be identified by employing classical detectors to examine any unauthorized signals (i.e. light originating from the attack that enter the sender's system).

Security proofs demonstrated that QKD protocols like BB84 are theoretically secure against a wide range of attacks permitted by quantum mechanics, even in scenarios where adversaries possess unlimited resources, including both classical and quantum computing power. These proofs, known as unconditional security proofs, guarantee that the protocol remains secure regardless of the eavesdropper's available resources. However, certain conditions must be met for these proofs to hold, such as ensuring that the adversary cannot physically access the encoding and decoding devices used by the communicating parties, using trusted and truly random number generators, authenticating the classical communication channel with an unconditionally secure authentication scheme, and encrypting the message using a scheme similar to a one-time pad.

7.7   QKD DISAPPROVAL BY GOVERNMENTAL INSTITUTIONS

As cybersecurity and quantum technologies continue to evolve, approaches to providing security in this changing landscape are coming under scrutiny from governmental institutions. One such method that is facing deprecation is QKD. This revolutionary technology is being scrutinized for its effectiveness in keeping up with the advancements in cyberattacks. While QKD was once regarded as the gold standard in cryptography, concerns were raised about its vulnerability to certain types of attacks. As a result, governmental institutions are now exploring alternative secure communication methods that can better withstand sophisticated attacks. In this section, we examine the reasons behind the deprecation of QKD by governmental institutions and explore the alternative encryption technologies that are being considered.

While QKD promises unparalleled security guarantees, its practical implementation presents several challenges. One of the main challenges is the need for dedicated infrastructure to support QKD networks. QKD relies on the transmission of quantum states through optical fibres, which are prone to losses over long distances. This limitation restricts the deployment of QKD to relatively short communication links, rendering it impractical for widespread use. Moreover, the cost associated with implementing QKD networks is a significant obstacle for many governmental institutions. The specialized hardware and equipment required for QKD can be excessively expensive, making it challenging to rationalize the investment, especially when alternative encryption methods are available at a lower cost. The complexity of QKD technology and the need for skilled personnel to maintain and operate QKD networks further add to the challenges of implementation. The specialized skills and expertise required to deploy and maintain QKD systems may not be readily available in governmental institutions, making it difficult to adopt QKD on a large scale.

In recent years, governmental institutions and critical infrastructure providers began to deprecate the use of QKD as their primary secure communication method. Growing concerns about its practicality and vulnerability to certain types of attacks led to a shift towards alternative encryption technologies. The deprecation of QKD by governmental institutions does not indicate a complete rejection of quantum encryption principles. However, it demonstrates a recognition of the evolving threat landscape and the need for more adaptable and robust encryption solutions.

Several reasons contribute to the deprecation of QKD by governmental institutions. One of the primary concerns is the vulnerability of QKD to side-channel attacks. While QKD's encryption keys are theoretically secure, risks can arise from the implementation of the hardware and software components that support the QKD infrastructure. Side-channel attacks were experimentally verified and demonstrated as a means of compromising the security of the communication channel.

Another reason for the deprecation of QKD is the fast advancement of quantum computing. Quantum computers have the potential to break the encryption algorithms used in traditional and quantum-resistant encryption methods, including QKD. With the increasing power and availability of quantum computers, the security provided by QKD may no longer be adequate to protect against future threats.

Additionally, the practical limitations of QKD (e.g. the transmission distance restrictions and the cost of implementation) contributed to its deprecation. Governmental institutions require encryption methods that can be implemented at a larger scale, span longer distances, and offer greater cost-efficiency. The intrinsic limitations of QKD make it challenging to meet these requirements, thus calling for the exploration of alternative encryption solutions. Governmental institutions and critical infrastructure providers started investigating alternative methods that can provide long-term security and scalability while addressing the limitations of QKD.

The US National Security Agency (NSA) does not recommend the usage of QKD and considers PQC or quantum-resistant cryptography as a more cost-effective and easier to maintain than QKD [7]. NSA does not recommend the adoption of QKD and does not anticipate certifying or approving any QKD for usage in National Security Systems (NSS) unless these limitations are overcome.

1. Partial solution: QKD provides keying material for encryption but cannot authenticate transmission sources, requiring additional cryptographic mechanisms.
2. Specialized equipment: QKD relies on physical properties requiring dedicated fibre connections or free-space transmitters. This makes QKD integration into existing networks challenging.
3. Increased costs and insider threats: Implementing QKD networks brings additional infrastructure costs and introduces risks from trusted relays, including insider threats.
4. Challenges in security validation: The process of validating QKD security is complex due to the difference between theoretical and practical security, compounded by hardware vulnerabilities and small error tolerances.
5. Denial of service risk: QKD's sensitivity to eavesdropping implies a significant risk of denial of service attacks.

Similarly, a European Union Agency for Cybersecurity of EU (ENISA) briefing paper on QKD possibilities, limitations, and issues in NSS cites several important areas of uncertainty and debate in QKD, including [8]:

1. The cost and complexity of QKD implementation: The security of QKD depends on its ability to exchange very large keys to provide forward security. The question arises as to when it becomes "worthwhile" to invest in QKD equipment rather than opting for alternative methods such as the exchange of very large keys; one such solution that stores keys on hard drives is [9].
2. The concept of unconditional security depends on a set of incomplete assumptions. All attacks on QKD depend on flawed modelling of theory in implementation caused by hidden assumptions. It is "impossible" for any implementation to be completely resistant to the discovery of a new assumption which was not previously considered.
3. Concerns that having one link in the security chain that is strong, when the other links are weaker.

In 2016, the UK National Cyber Security Centre (NCSC) published a white paper advocating for the research into developing PQC as a more practical and cost-effective approach, rather than QKD, to protecting real-world communication systems from the threat of quantum computers [10]. NCSC followed this up with a report in 2020 reaffirming their stance that they do not endorse the use of QKD for any government or military applications, citing the "specialised hardware requirements of QKD" as a key reason [11].

In January 2024, the French Cybersecurity Agency (ANSSI), Federal Office for Information Security (BSI), Netherlands National Communications Security Agency (NLNCSA), Swedish National Communications Security Authority, and Swedish Armed Forces published a technical report on the uses and limits of QKD [12]. This report concludes that QKD is not yet sufficiently mature from a security perspective and "a clear priority should be given to the deployment of PQC" which is not very different from existing cryptography than QKD. It also highlights the limitations of QKD, with specific reference to its need for specialized communication infrastructure. The report discusses other QKD functional limitations and its applicability to only certain niche use cases.

Most government institutions recommended using PQC as an alternative to QKD. The NSA report [7] concludes that QKD and PQC vendors often make strong claims based on published theory, e.g. that this technology offers "guaranteed" security based on the laws of physics. However, communications and security requirements physically conflict with using QKD and PQC. Achieving a balance between these two fundamental aspects necessitates meticulous engineering with minimal tolerance for error. Thus, the security of QKD and PQC is primarily determined by the specific implementation rather than being guaranteed by the laws of physics.

7.8 PQC AS AN ALTERNATIVE FOR QKD

The deprecation of QKD motivated the exploration and development of alternative encryption technologies that can effectively tackle the changing security challenges encountered by governmental institutions. Several approaches are being considered as potential replacements for QKD, each with its own strengths and limitations.

One such alternative is PQC. Both QKD and PQC rely on physics principles and employ similar technology to communicate over a dedicated communications link. It is widely accepted that physics principles allow QKD or PQC to detect the presence of an eavesdropper, a feature not supported in classical cryptography. PQC focuses on developing encryption algorithms that are resistant to attacks from both classical and quantum computers. By leveraging mathematical problems that are computationally hard to solve even with quantum computing, PQC has the potential to offer long-term security in a post-quantum era. Extensive research and standardization efforts are underway to identify and promote PQC algorithms that can serve as the next generation of encryption [13]. The NIST is currently engaged in a rigorous selection process to identify quantum-resistant (or post-quantum) algorithms for standardization (with three PQC alogrithms FIPS 203, 204 and 205 published on 13 August 2024).

Another viable alternative is the development of quantum-resistant encryption methods. These encryption techniques are designed to protect against attacks from both classical and quantum computers without relying on the unique properties of quantum mechanics. By employing mathematical algorithms that are resistant to quantum attacks, quantum-resistant encryption methods present a feasible solution for secure communication in a post-quantum world.

Table 7.2
A Comparison between QKD and PQC

Aspect	QKD	PQC
Underlying principles	Relies on the principles of quantum mechanics	- Relies on classical cryptographic algorithms - Provides security guarantee as no method exists to break the PQC algorithm
Security guarantees	- Offers information-theoretic security guarantees based on the laws of physics - Special hardware required	- Offers security guarantees based on mathematical problems that are believed to be hard for both classical and quantum computers - Most implementations are software-only, no special hardware is required
Vulnerability to quantum computers	- Immune to attacks from quantum computers - Cannot be successfully intercepted without detection	Vulnerable to attacks from quantum computers, but resistant to attacks from classical computers
Key exchange mechanism	Generates cryptographic keys using the exchange of quantum states between parties	Uses mathematical algorithms for key exchange and encryption/decryption
Practical implementations	- Limited by current technological constraints, with implementations primarily in research and niche applications - Fibre optics or free space communication is required	- Widely deployed and integrated into existing cryptographic systems, with practical implementations in various domains - Applicable to all types of traditional digital communication media, e.g. RF, wired network, and optical communication
Key management	Provides provably secure key distribution, but may require specialised infrastructure and protocols	Requires ongoing research and development to ensure resilience against emerging threats and advancements in quantum computing
Performance	Offers relatively low key generation rates and limited transmission distances	Offers faster key generation rates and greater flexibility in terms of key size and transmission distance
Resilience to attacks	Resilient against quantum attacks but vulnerable to certain side-channel attacks	Resilient against classical attacks but potentially vulnerable to quantum attacks once large-scale quantum computers become available
Cost	Relatively high cost due to the need for specialised hardware and new communication infrastructure	Relatively low cost based on software
Repeater compatibility	A repeater can receive a quantum channel, decode it into classic bits, and re-encrypt and retransmit it to another quantum channel	Fully compatible with existing digital repeater technology
Mobile device compatibility	Limited to line-of-sight nodes only	Compatible with all types of communication supported by mobile devices
Digital signature compatibility	Generally not used for digital signature applications due to unique properties	Compatible with digital signature schemes for authentication and non-repudiation purposes

Additionally, advancements in homomorphic encryption [14], multi-party computation [15], and secure multi-party communication protocols [16] are being explored as potential alternatives to QKD. These encryption methods prioritize protecting data at rest, in transit, and during processing, offering a comprehensive approach to secure communication within governmental institutions.

The transition away from QKD involves not only the adoption of alternative encryption technologies but also the need to update existing infrastructure, retrain personnel, and establish new security protocols. One of the core implications of QKD deprecation is the need to reevaluate the security posture of governmental institutions. The vulnerabilities and limitations of QKD are to be addressed by adopting more robust encryption methods. This reevaluation includes assessing current communication networks, identifying potential vulnerabilities, and implementing adequate measures to reduce risks. Furthermore, the deprecation of QKD emphasises the importance of continuous research and development in the field of encryption. As the threat landscape changes, encryption methods must adapt to tackle new challenges. Governmental institutions should allocate resources to research to identify and develop encryption technologies that are resilient against future attacks and provide long-term security.

7.9 CONCLUSION

In conclusion, the security analysis of QKD reveals both its strengths and limitations. While QKD offers information-theoretic security guarantees against quantum adversaries, it remains susceptible to many practical attacks, including eavesdropping, Trojan horse attacks and denial-of-service attacks. Additionally, the requirement for specialized infrastructure and the vulnerability to side-channel threats pose significant challenges to the widespread adoption and practical implementation of QKD systems. Although published theories suggest that QKD technology is theoretically secure, its current state is characterized by its immaturity, limited scalability, and susceptibility to practical attacks.

In light of these considerations, it is evident that while QKD holds promise and theoretically offers "guaranteed" security for key exchange, its immaturity and practical limitations hinder its widespread adoption. Post-quantum cryptography emerges as a more feasible and practical alternative, offering mature cryptographic algorithms that are resistant to both classical and quantum attacks. As organizations navigate the landscape of cryptographic security, they must carefully weigh the trade-offs between theoretical security guarantees and practical considerations, recognising that PQC offers a more robust and flexible solution in the face of growing threats and technological advancements.

REFERENCES

1. Antonio Acin, Nicolas Gisin, and Valerio Scarani. Coherent-pulse implementations of quantum cryptography protocols resistant to photon-number-splitting attacks. *Phys. Rev. A*, 69(1):012309, 2004.

2. Bing Qi, Chi-Hang Fred Fung, Hoi-Kwong Lo, and Xiongfeng Ma. Time-shift attack in practical quantum cryptosystems. *Quantum Info. Comput.*, 7(1):73–82, 2007.

3. Vadim Makarov, Andrey Anisimov, and Johannes Skaar. Effects of detector efficiency mismatch on security of quantum cryptosystems. *Phys. Rev. A*, 74:022313, 2006.

4. J. B. Altepeter, D. Branning, E. Jeffrey, T. C. Wei, P. G. Kwiat, R. T. Thew, J. L. O'Brien, M. A. Nielsen, and A. G. White. Ancilla-assisted quantum process tomography. *Phys. Rev. Lett.*, 90:193601, 2003.

5. Pablo Arteaga-Díaz, Daniel Cano, and Veronica Fernandez. Practical side-channel attack on free-space qkd systems with misaligned sources and countermeasures. *IEEE Access*, 10:82697–82705, 2022.

6. Nitin Jain, Elena Anisimova, Imran Khan, Vadim Makarov, Christoph Marquardt, and Gerd Leuchs. Trojan-horse attacks threaten the security of practical quantum cryptography. *New J. Phys.*, 16(12):123030, 2014.

7. National Security Agency/Central Security Service, Cybersecurity, and Quantum Key Distribution (QKD), and Quantum Cryptography QC – nsa.gov. `https://www.nsa.gov/Cybersecurity/Quantum-Key-Distribution-QKD-and-Quantum-Cryptography-QC/`. [Accessed 23-03-2024].

8. Giles Hogben. Enisa briefing: Quantum key distribution. Technical report, European Network and Information Security Agency, 2009.

9. Christopher Mobley. Electronic anti-tampering device, February 2020.

10. NCSC. Quantum security technologies. Technical report, National Cyber Security Centre, November 2016.

11. NCSC. Quantum security technologies. Technical report, National Cyber Security Centre, March 2020.

12. NLNCSA SNCSA SAF ANSSI, BSI. Position paper on quantum key distribution. Technical report, French Cybersecurity Agency, Federal Office for Information Security, Netherlands National Communications Security Agency, Swedish National Communications Security Authority, Swedish Armed Forces, January 2024.

13. Post-Quantum Cryptography – CSRC – CSRC – csrc.nist.gov. `https://csrc.nist.gov/projects/post-quantum-cryptography`. [Accessed 22-03-2024].

14. Mostefa Kara, Abdelkader Laouid, Mohammed Amine Yagoub, Reinhardt Euler, Saci Medileh, Mohammad Hammoudeh, Amna Eleyan, and Ahcène Bounceur. A fully homomorphic encryption based on magic number fragmentation and el-gamal encryption: Smart healthcare use case. *Expert Syst.*, 39(5):e12767, 2022.

15. Christian Vincent Mouchet. Multiparty homomorphic encryption: From theory to practice. Technical report, EPFL, 2023.

16. Mostefa Kara, Abdelkader Laouid, Ahcène Bounceur, Mohammad Hammoudeh, and Muath AlShaikh. Perfect confidentiality through unconditionally secure homomorphic encryption using otp with a single pre-shared key. *J. Inform. Sci. Eng.*, 39(1), 2023.

8 Quantum Random Number Generation

Christoph Capellaro and Mohammad Hammoudeh

8.1 INTRODUCTION

Random numbers have a surprisingly wide field of application. Depending on the specific use case, the requirements on their quality vary, of course. It is kind of obvious that due to its probabilistic behaviour, quantum technology is a natural area of investigation when looking at appropriate sources for random numbers.

In the following, we provide an overview of application areas where random numbers are important for a variety of reasons:

- Cryptography: One of the most critical applications of real random numbers is in cryptography. Random numbers are used to generate keys for encryption and decryption, ensuring the security of communication. A lack of true randomness can make cryptographic systems vulnerable to attacks because predictability allows malicious actors to potentially guess or determine key values.
- Statistical sampling: In statistics, random numbers are used for random sampling and randomised controlled trials. This ensures that samples are representative of a population, and that the results of trials are not biased by any underlying patterns.
- Simulations: Complex systems and physical phenomena are often simulated using random numbers to model random variables and stochastic processes. These simulations can be used for weather prediction, financial market forecasting, and scientific research.
- Gaming: In computer games, random numbers are necessary to ensure fairness and unpredictability. For instance, dice rolls, shuffles of a deck of cards, or random events within a video game rely on random numbers to provide an element of chance.
- Randomised algorithms: Many algorithms, particularly those in computer science, use randomness to optimise performance. For example, randomised QuickSort or randomised algorithms for primality testing take advantage of random numbers to improve average-case performance.
- Art and music: Random numbers can be used to generate procedural content, such as textures, landscapes in video games, or even music. This

DOI: 10.1201/9781003475286-8

allows for the creation of unique and diverse content without the need for explicit human design.

Typically, nowadays computers use Pseudo-Random Number Generators (PRNG) using some seed that is then shuffled by a certain algorithm. The seed could be any given parameter available to the computer, like the current time, the amount of storage used, CPU utilisation, etc. Pseudo-random number generation has drawbacks, as the seed could be predictable, and the generated random numbers will exhibit patterns over long sequences or be subject to deterministic behaviour. Hence, the importance of random numbers being real lies in their unpredictability and the inability to reproduce the same sequence, which is crucial for the integrity and reliability of the processes and systems that rely on them.

For a more systematic approach, we discuss typical problems of common random number generators:

- Predictability: Since PRNGs use deterministic algorithms, given the initial seed, the sequence of numbers can be predicted. This predictability is a major issue for applications like cryptography where unpredictability is paramount.
- Periodicity: PRNGs have a finite period after which they repeat the same sequence of numbers. For applications with large random number requirements or long-running processes, this repetition can become problematic.
- Bias and non-uniformity: Some PRNGs may produce numbers that are not evenly distributed across the expected range. This can skew results in statistical simulations or other processes that rely on a uniform distribution of random numbers.
- Correlation: Numbers produced by PRNGs can be correlated in complex ways, which might affect statistical properties of the sequence. In simulations or statistical analyses, such correlations can lead to inaccurate results.
- Seeding issues: The quality of the random number sequence is also dependent on the quality of the seed. Poor seeding can lead to reduced randomness and security vulnerabilities in cryptographic systems.
- Algorithmic complexity: Some PRNGs may be complex to implement or require significant computational resources, which can be inefficient for certain applications.
- Not suitable for high-stakes security: Since PRNGs are not truly random, they are not suitable for high stakes security applications. Secure systems require Cryptographically Secure Pseudo-Random Number Generators (CSPRNGs) that resist prediction even if some of the initial conditions or internal state become known to an attacker.

To address these problems, CSPRNGs, Hardware Random Number Generators (HRNG), or True Random Number Generators (TRNG) are used where higher-quality randomness is needed. CSPRNGs are designed to provide a higher level of security, making their output computationally infeasible to predict, while TRNGs derive randomness from physical processes, e.g., electronic noise, which are inherently unpredictable. TRNGs often suffer from slower generation rates, and hardware components may degrade over time, impacting the quality of randomness. Additionally, there remains the question of whether these physical processes could, in theory, be predicted or manipulated.

8.2 QUANTUM TECHNOLOGY FOR RANDOM NUMBER GENERATION

Quantum technology can be used for random number generation by exploiting the inherent randomness of quantum mechanical processes. Unlike classical physics, which is deterministic in nature, quantum mechanics involves intrinsic probabilities that lead to true randomness. Several quantum phenomena have been harnessed to develop Quantum Random Number Generators (QRNGs):

- Quantum superposition: One fundamental aspect of quantum mechanics is the principle of superposition, where a quantum system can be in a combination of all its possible states simultaneously. When such a superposed state is measured, the outcome is probabilistic, collapsing to one of the possible states with certain probabilities. This unpredictability can be used to generate random numbers.
- Photonic emission: A common approach involves measuring the randomness in the number of photons emitted by a light source over a certain period of time. Due to quantum uncertainty, the emission of photons is a random process that can be used to generate random numbers.
- Single photon detection: Utilising single-photon detectors and a beam splitter, photons are sent one at a time towards the splitter, and which path they take after the split is intrinsically random. Recording the path each photon takes—either transmitted or reflected—can be directly translated into random bits.
- Vacuum fluctuations: The quantum vacuum is subject to fluctuations due to the uncertainty principle, leading to the temporary emergence of particle-antiparticle pairs. Measuring these fluctuations can yield a source of quantum randomness that can be converted into random numbers.

QRNGs have the advantage of true randomness, as quantum processes are non-deterministic in nature, and potentially high-speed generation, which is valuable for applications requiring large quantities of random data. Despite these advantages, QRNGs can be complex and expensive to implement and may require careful calibration and maintenance to ensure that the underlying quantum processes are not influenced by environmental factors or equipment biases.

8.3 CURRENT STATE OF TECHNOLOGY

The approaches to leverage the fundamental randomness inherent in quantum mechanical systems are subject to ongoing research and development to optimise their efficiency, reliability, and scalability for practical applications. In this section, we give an overview of the current state-of-the-art of QRNG technology.

A QRNG and a photon generator for a QRNG are presented in [7]. To emit photons, the photon generator is operated in a spontaneous mode below a lasting threshold. The QRNG monitors photons emitted by the photon generator that possess at least one random characteristic to generate a random number. A photon emitter and an amplifier coupled to the photon emitter may be included in one embodiment of the photon generator. Multiplexing multiple random numbers may be made possible by the amplifier, which may also permit the photon generator to be utilised in the QRNG without introducing a substantial amount of bias into the random number. In the spontaneous mode, the amplifier may also desensitise the photon generator to variations in the power supplied to it. A tapered diode amplifier may serve as both the photon emitter and amplifier in one embodiment. In the spontaneous mode, the amplifier may also desensitise the photon generator to variations in power supplied to it. A tapered diode amplifier may serve as both the photon emitter and amplifier in one embodiment.

National Institute of Standards and Technology (NIST) researchers have developed a new technique, detailed in Nature [1], which utilises quantum mechanics to generate numbers that are assuredly random. This quantum-based method, which generates digital bits using photon measurements, enhances the randomness beyond what previous methods could guarantee. The core concept involves using entangled photon pairs generated from an intense laser hitting a special crystal. By measuring these photons, researchers produce truly random numbers.

NIST achieves this by conducting a Bell test experiment, generating a long string of bits through measuring properties of photon pairs and processing these bits into a more uniform and concentrated string of random bits, ensuring each bit has an equal chance of being 0 or 1. From over 55 million trials, they extracted 1,024 bits with a very high degree of uniformity. This method is part of NIST's efforts to improve their public randomness beacon. The final random numbers generated are certified as truly random, even if the input conditions are known, provided the experiment itself is kept isolated.

There are many advantages of NIST's new quantum method for generating truly random numbers. The method's use of quantum mechanics ensures that the generated numbers are intrinsically unpredictable, making them ideal for cryptographic systems where the security depends on the randomness of keys. The protocol provides a level of certifiable randomness that surpasses classical methods.

Any implementation deviations from the theoretical model may impact the output randomness. To address this limitation, the authors of [6] present a source-independent approach for QRNG that produces certified randomness even when the source is uncharacterised and untrusted. The authors do not make any assumptions about the dimension of the sources in their randomness analysis. Their analysis considers the finite-key effect with the composable security definition. The input random seed has an exponentially shorter length than the output random bit. Experimental validation of this scheme shows that it achieves a high randomness generation rate by modifying a quantum key distribution system.

Main disadvantages of this new methodology are seen in following areas. The setup required for generating random numbers through Bell tests and photon measurements is more complex than classical methods, making it potentially more difficult and more expensive to implement. While the method produces highly random bits, the rate at which it can do so may be slower than traditional methods, potentially limiting its practicality for some applications that require high-speed random number generation. Quantum experiments like the one used in this method require specialised equipment, such as precise lasers, photon detectors, and cryogenic systems, that are resource-intensive in terms of cost and expertise. The original string of bits can be very long, and the randomness may be thinly spread within it. An extraction process is necessary to concentrate the randomness, which introduces an additional computational step. To ensure true randomness, the quantum system must be carefully calibrated and verified, which might require sophisticated knowledge and technology. The generated numbers can only be certified as random if the experimental setup is physically isolated from outside interference or hacking.

In theory effects in quantum physics are purely random. However, technical implementations do have their problems. Steinle et al. [2] mention that while optical devices have some advantages to other sources in the electromagnetic spectrum, there are still challenges to overcome when it comes to produce real random numbers. For example, with single photon detectors ambiguities can occur due to dead times, electrical jitter, and varying detection efficiencies. One way to overcome this is to deploy a post-processing, but this reduces the available output.

The approach discussed in [2] utilises so-called degenerated or two-state outcome Optical Parametric Oscillators (OPOs). Their source of randomness is the generation of single photons in a spontaneous down conversion process caused by pumping a non-linear gain crystal. A defect in the crystal is excited by a laser which makes the crystal emitting single photons with random phases. The set-up described by the authors requires little calibration before and during the process of random number generation and hence produces random numbers with high frequency. In this chapter from 2017, they think that a frequency of 1 Mbps for QRNG is possible. Latest results with comparable technology on an on-chip solution already reach several Gbps [3]. Physical random numbers were generated at 500 Mbps using a high-speed balanced photodetector to quantify the quadrature amplitudes of vacuum states [8].

The authors of [9] describe the generation of quantum random numbers at multi-Gbps rates. This QRNG integrates real-time randomness extraction to produce very high-purity random numbers based on quantum events at most tens of nanoseconds in the past. The system fulfils the rigorous requirements of quantum non-locality tests designed to eradicate the timing loophole. This QRNG is described through spontaneous-emission-driven phase diffusion in a semiconductor laser, digitisation, and extraction by parity calculation using multi-GHz logic chips. Thorough experimental proof of the quality of the random numbers and analysis of the randomness extraction are given in [9]. Contrary to the prevalent models of randomness generators discussed in the literature, the authors contend that spontaneous emission randomness can be derived from a singular source.

An unbiased QRNG whose statistical properties can be arbitrarily programmed without the need for any postprocessing is presented in [10]. This method involves measuring the arrival time of single photons in shaped temporal modes that are customised with an electro-optical modulator. Quantum random numbers are created directly in customised probability distributions. These quantum random numbers satisfy all randomness tests of the NIST and Dieharder test suites without any randomness extraction. The min-entropies of the generated random numbers are measured close to the theoretical limits, suggesting their near-ideal statistics and ultrahigh purity. This technique is highly adaptable and can be implemented arbitrarily; it has the potential to be utilised in a wide range of data analysis domains.

Not in every instance a user is interested in purchasing specific equipment to have a reliable source of true or certified random numbers. Online services can be an alternative in this case. Miszczak [4] discusses the deployment of an online service for QRNG. This is the backbone of the Truly Random Quantum States (TRQS) package available for the Mathematica computing system. In his paper the author provides statistics showing that this service could provide random real numbers in the range from [0,1] of size 10^7 within a second. The functionality to utilize the online QRNG was incorporated into the latest version of the package, which also includes new functions for retrieving lists of random numbers. The latest version offers faster access to high-quality sources of random numbers making it suitable for simulations requiring large amounts of random data. This outperforms USB devices that can be used for the same purpose. Of course, the performance depends on network bandwidth and latency.

An Italian research group describes a design of a high-performance source-device independent Semi-Device Independent (Semi-DI) QRNG using an Photonic Integrated Chip (PIC) [5]. The authors discuss the need for evaluating non-idealities in quantum-based RNGs, as these imperfections can leak information and compromise security. They classify QRNGs as Fully Trusted (FT), which are simple with high generation rates but low security; Device Independent (DI), which offer high security but are difficult to implement and have low rates; and Semi-DI, which provide a balance between security and speed by trusting only specific aspects of the device. The Semi-DI QRNG protocol divides the QRNG into a quantum source and a measuring apparatus, with the latter being trusted and fully characterised.

Design, fabrication, and packaging of a PIC uses semiconductor technology to realise a heterodyne receiver for the QRNG. This design reduces power consumption compared to traditional optical devices. A series of tests performed in [5] confirmed the chip's high phase accuracy and low losses, indicating its potential for high-speed QKD, especially in space applications where power and size are limited. In an experimental setup, the device demonstrated a secure generation rate of over 20 Gbps, making it suitable for high-speed QKD. The QRNG's security is strengthened by its design and robustness against environmental challenges. The paper concludes that the developed QRNG combines high security, performance, and low power consumption in a compact device, making it an ideal solution for demanding applications such as ground and space-based QKD.

8.4   VACUUM FLUCTUATIONS

Quantum vacuum fluctuations are intrinsic and unpredictable variations that occur in the quantum field at the smallest scales of space and time, even in a complete vacuum. This phenomenon is a result of the Heisenberg uncertainty principle, which states that it is impossible to simultaneously know with infinite precision certain pairs of physical properties, such as the position and momentum of a particle. In the context of the vacuum, the uncertainty principle predicts fluctuations of energy and hence the transient appearance and disappearance of particle-antiparticle pairs, known as "virtual particles." These fluctuations are truly random, as they do not follow any predetermined pattern and cannot be predicted ahead of time.

In the creation of random number generators based on quantum vacuum fluctuations, this inherent randomness is harnessed to generate random numbers. Therefore, an empty quantum state or vacuum state is prepared. In this state, quantum fluctuations occur naturally due to the uncertainty principle. The fluctuations measured using, e.g. a high-speed photodetector can also measure fluctuations in the electromagnetic field of the vacuum. A common method for detecting vacuum fluctuations is homodyne detection. A strong "local oscillator" laser beam (which provides a reference phase) is combined with the vacuum state. The resulting signal mixes vacuum fluctuations with the reference beam, which leads to a measurable signal that still reflects the randomness of the quantum vacuum. Heterodyne detection is a similar method to homodyne detection, but it measures two non-commuting variables (like position and momentum, or two orthogonal quadrature phases of the light field) simultaneously, increasing the randomness captured in the measurement.

Then the physical measurements, which could be voltage fluctuations or light intensity differences over time, are converted into a digital form. This conversion uses a high-speed analogue-to-digital converter to represent the measurements as digital values. The raw digital data often undergoes post-processing to enhance randomness and remove biases, often using statistical techniques or cryptographic hash functions. This step ensures that the output numbers pass rigorous statistical tests for randomness. The final outcome of this process is a sequence of random numbers generated from the fundamental quantum fluctuations of the vacuum state.

In [11], the authors propose an optical randomness generator that uses the bistable output of an OPO. This QRNG does not consider detector noise, and post-processing (e.g., randomness distillation or distribution transformation) is minimal. Upon entering the bistable domain, the initial phase of the output is determined by fluctuations in the vacuum. Then, the phase is rigidly locked and can be precisely determined against a pulse train generated by the pump laser. Additional verification of the derived binary outcome's randomness is provided through an examination of the generated conditional entropies.

8.5   ATOMIC DECAY

Random numbers can also be generated from atomic decay, another quantum mechanical process that is fundamentally random and unpredictable. Atomic decay refers to the process where an unstable atomic nucleus releases energy by emitting radiation as it changes (or decays) into a more stable form. This radioactive decay occurs spontaneously and is independent of any external conditions, such as temperature or pressure, and follows a characteristic decay law that is probabilistic in nature.

A suitable radioactive isotope is chosen, one that decays by emitting particles or radiation at a rate appropriate for the intended application. Common isotopes used include Americium-241 or Cesium-137. A detection device, such as a Geiger-Müller tube or scintillation counter, is positioned near the radioactive source. This detector is sensitive to the high-energy particles or photons (e.g. alpha particles, beta particles, gamma rays) emitted during the decay. When a particle is emitted due to decay, it interacts with the detector, typically by ionising the gas within a Geiger tube or exciting a scintillator. This interaction generates an electrical pulse. The electrical pulses from the detection device are counted and timestamped. This can be done using electronic counters or computer interfaces designed to handle signals from radioactive decay. Random Number Generation: Random numbers are derived from the occurrences of decays in one of several ways: The random numbers can be generated based on the varying intervals between consecutive decays. Due to the stochastic nature of radioactive decay, these intervals are random and can be used directly as random data or further processed to fit a desired numerical range or distribution. Another approach is to count the number of decays within a fixed time window, using the count as a source of randomness. Since the timings or counts may not be uniformly distributed or may have some predictable structure due to the decay properties of the isotope, the raw data may undergo post-processing. This could include bit extraction, hashing, or other statistical methods to ensure uniformity and remove biases to produce high-quality random numbers. After post-processing, the extracted data serve as a sequence of random numbers.

The unpredictability of quantum processes such as radioactive decay makes atomic decay an excellent source of true randomness. However, several factors must be considered when generating random numbers using this method, such as the half-life of the isotope, the impact of background radiation, the efficiency and speed of the detection apparatus, and the integration and shielding for practical and safe application.

8.6 CONCLUSION

The technology required to produce quantum-based random numbers is quite advanced and hence various products are already on the market. There are several companies that offer random number generators based on quantum technology, e.g.:

- ID Quantique: This Switzerland-based company was one of the pioneers in commercial QRNGs with its Quantis product line. These devices use photonics to generate randomness, specifically through the quantum property of photon arrival times.

- QuintessenceLabs: An Australian company that produces high-speed QRNGs as part of their cybersecurity solutions. Their qStream QRNG harnesses the unpredictability of quantum tunneling where electrons randomly overcome an energy barrier, producing a stream of true random numbers.
- Quside: A spin-off from the Institute of Photonic Sciences in Spain, offering QRNG products for various applications. Their FMC 400 is a QRNG module that can be integrated into larger systems for providing quantum randomness.
- QuantumCTek: This Chinese company offers QRNGs that are designed to be used in various applications such as secure communications and cryptographic protocols.
- Random Quantum: A company providing a quantum random number generator called RQubit that uses quantum tunneling phenomena to generate random numbers.
- Infineon: The German semiconductor manufacturer collaborated with the quantum technology company ID Quantique to create a QRNG chip that can be embedded in mobile devices.

These products range from components like chips and modules that can be integrated into larger system, to stand-alone appliances designed for specific uses such as secure communications, encryption keys generation, and scientific research. Additionally, there are QRNGs available as a service, where users can access quantum-generated random numbers via an API, often provided by those same companies.

As QRNG technology continues to mature, it's expected that more products and services will become available, catering to an increasing demand for high-quality randomness in security, scientific research, and various industries that rely on unpredictability and entropy.

REFERENCES

1. P. Bierhorst, E. Knill, S. Glancy, Y. Zhang, A. Mink, S. Jordan, A. Rommal, Y.-K. Liu, B. Christensen, S.W. Nam, M. Stevens, L.K. Shalm. Experimentally Generated Random Numbers Certified by the Impossibility of Superluminal Signaling. Nature, April 12, 2018.

2. Tobias Steinle, Johannes N. Greiner, Jörg Wrachtrup, Harald Giessen, Ilja Gerhardt. Unbiased All-Optical Random-Number Generator. Physical Review X, vol. 7, no. 4, id.041050, 2017.

3. Muhammad Imran, Vito Sorianello, Francesco Fresi, Luca Potì, Marco Romagnoli. Quantum Random Number Generator Based on Phase Diffusion in Lasers Using an On-chip Tunable SOI Unbalanced Mach-Zehnder Interferometer (uMZI). Optical Fiber Communication Conference (OFC), 2020. https://doi.org/10.1364/ofc.2020.m1d.5.

4. Jaroslaw Adam Miszczak. Employing Online Quantum Random Number Generators for Generating Truly Random Quantum States in Mathematica. arXiv:1208.3970, 2013.

5. Tommaso Bertapelle et. al. High-Speed Source-Device-Independent Quantum Random Number Generator on a Chip. 2023. https://doi.org/10.48550/arXiv.2305.12472.

6. Z. Cao, H. Zhou, X. Yuan, X. Ma. Source-Independent Quantum Random Number Generation. Physical Review X, vol. 6, no. 1, APS, 2016. https://doi.org/10.1103/PhysRevX.6.011020.

7. Raphael C. Pooser. Quantum Random Number Generator. United States, 2016. https://www.osti.gov/servlets/purl/1252208.

8. K. Hirakawa. Generation of Physical Random Numbers by Using Homodyne Detection. Quantum Information Science and Technology II, vol. 9996, 2016. https://doi.org/10.1117/12.2241362.

9. M. Mitchell, C. Abellan, W. Amaya. Quantum Random Number Generation for Loophole-Free Bell Tests. vol. 2015, 2015. https://meetings.aps.org/Meeting/DAMOP15/Session/G7.8

10. L. Nguyen, P. Rehain, Y. M. Sua, Y. P. Huang. Programmable quantum random number generator without postprocessing. Optics Letters, vol. 43, no. 4, 631–634, 2018. https://doi.org/10.1364/OL.43.000631.

11. T. Steinle, J. N. Greiner, J. Wrachtrup, H. Giessen, I. Gerhardt. Unbiased All-Optical Random-Number Generator. Physical Review X, vol. 7, no. 4, APS, 2017. https://doi.org/10.1103/PhysRevX.7.041050.

9 Managing the Quantum Cybersecurity Threat
Harvest Now, Decrypt Later

Harbaksh Singh

9.1 INTRODUCTION

From simulations in the pharmaceutical industry that could lead to drug discoveries, optimization problems in logistics and supply chains, to new techniques for machine learning, the potential applications of quantum computing could revolutionize various industries. However, the flip side to this exciting innovation is the threat that quantum computing poses to cybersecurity. A sufficiently powerful quantum computer could break today's public key encryption within seconds. In the hands of ill-intentioned individuals or states, this capability can disrupt systems that heavily rely on encryption for security, such as financial systems, communication networks, and national defense infrastructures. Therefore, organizations and governments worldwide have started to take the quantum threat seriously and are investing in quantum-resistant cryptographic technologies and preparing for the "Y2K moment" of the quantum era.

In the swiftly evolving landscape of cybersecurity, quantum computing presents a double-edged sword; it holds the potential for groundbreaking advances in data processing capabilities while simultaneously posing a significant threat to current cryptographic defenses. The phrase "Harvest Now, Decrypt Later" captures a particularly concerning strategy that adversaries may employ in anticipation of quantum advances. By this method, malicious actors collect encrypted data with the intent to decrypt it in the future, when quantum technology may render traditional encryption obsolete. Such a scenario necessitates a proactive and strategic approach to managing quantum cybersecurity risks. Governments, organizations, and cybersecurity experts are now faced with the urgent task of developing quantum-resistant cryptographic methods to protect sensitive information not just for today, but for the years to come, thereby ensuring that the data harvested now remains secure against the decrypting powers of future quantum computing.

DOI: 10.1201/9781003475286-9

9.2   QUANTUM CRYPTOGRAPHY

Quantum cryptography is a technique that applies the unique properties of quantum mechanics to secure communication. The most well-known technique is Quantum Key Distribution (QKD), which provides an exceedingly secure process for two parties to produce a shared random secret key known only to them, which can be used to encrypt and decrypt confidential messages. You may recall that this topic was covered in detail in previous chapters.

The fundamental principle that gives QKD its ultra-secure nature is the ***Heisenberg Uncertainty Principle***, which states that you cannot simultaneously measure two complementary variables (like position and momentum) of a quantum system without disturbing the system. What this means in the context of QKD is that any attempt by an eavesdropper to intercept the key will unavoidably alter its state, and this alteration can be detected by the legitimate parties in communication, alerting them to the presence of the eavesdropper.

While QKD offers theoretically unbreakable security, it is not entirely free from practical challenges. These include transmission limitations due to fiber optic cable loss and the technological sophistication required for implementation. Despite these limitations, however, QKD has been successfully deployed in certain contexts. As an example, in 2017 China's Micius satellite, the world's first quantum satellite, successfully used QKD for a secure video call between Beijing and Vienna [1]. The development of quantum communication networks, such as the quantum internet, could pave the way for broader deployment of quantum cryptography and help secure communications against the quantum threat.

Quantum algorithms capitalize on qubits' properties of superposition and entanglement. Two prime examples of such algorithms are Shor's and Grover's algorithms, both of which capitalize on the advantages of quantum mechanics. While Shor's algorithm provides the ability to use quantum computing to solve the factorization/discrete logarithm problem in polynomial time, Grover's algorithm allows for a quadratic acceleration of a search in an unsorted database, which puts symmetric encryption systems at the risk of brute force attacks. These topics are covered in detail in previous chapters.

In summary, Shor's algorithm is well-known for its potential to disrupt traditional encryption methods. It can factor large integers into primes efficiently – a problem considered 'hard' for classical computers and the cornerstone of many of today's encryption methods, particularly public-key primitives such as RSA and EC-DSA. Shor's algorithm, on a sufficiently large quantum computer, could solve this problem exponentially faster than classical algorithms, thereby theoretically breaking RSA encryption.

Grover's algorithm, developed by Lov Grover of Bell Labs in 1996, is designed for searching an unsorted database or an unordered list. It offers a quadratic speedup compared to its classical counterparts. While not as damaging to today's cryptographic systems as Shor's algorithm, Grover's algorithm does impact the security of symmetric key cryptographic systems that many internet standards use, essentially halving the effective key length.

Research in quantum algorithms extends beyond the realms of Shor's and Grover's. IBM, Google, and other tech giants are making significant strides in developing new quantum algorithms for near-term quantum computers (Noisy Intermediate-Scale Quantum or "NISQ" devices [2]) that can perform useful tasks before full error-corrected quantum computers become a reality.

9.2.1 HASHES/KEYS (SYMMETRIC AND ASYMMETRIC)

Quantum computing's expected impact on cybersecurity primarily hinges on quantum algorithms' potential ability to rapidly decode cryptographic keys. The majority of modern cryptography systems rely on the mathematical complexity of certain problems (like factoring large primes) to ensure security, which a powerful quantum computer has the potential to shatter.

In symmetric key algorithms, the same cryptographic key is used for both encryption and decryption. Even so, these are expected to endure the advent of quantum computers relatively unscathed, albeit with a reduction in security level. For instance, AES-256, a commonly used symmetric key algorithm, is likely to become more vulnerable to decryption in a post-quantum world.

On the other hand, asymmetric or public key cryptography systems, which use different keys for encryption and decryption (public and private keys, respectively), are in danger. RSA and Elliptic Curve Cryptography (ECC), two popular asymmetric systems, could be broken by a sufficiently capable quantum computer leveraging Shor's algorithm as shown in Table 9.1.

It is fair to say that while the impact on symmetric cryptographic algorithms remains relatively contained, achieved through the adoption of longer keys or extended hash function outputs, public key cryptographic algorithms face a serious threat. This necessitates the replacement of current public key cryptographic algorithms and standards [3]. This imminent threat has galvanized the research and development of Post-Quantum Cryptography (PQC) cryptographic primitives resistant to quantum attacks. NIST, among others, are running programs to standardize quantum-resistant algorithms. Figure 9.1 provides an overview of selected cryptographic standards provided by NIST. PQC provides primitives that are resistant to quantum attacks which has several approaches [14]:

Lattice-based cryptography: This approach is based on the hardness of certain problems in lattice theory, such as the Shortest Vector Problem (SVP) and the Closest Vector Problem (CVP). Lattice-based systems like Learning With Errors (LWE) and Ring-LWE are currently seen as some of the most promising post-quantum systems.

Code-based cryptography: Code-based schemes rely on the difficulty of decoding a general linear code. The oldest post-quantum encryption algorithm, the McEliece cryptosystem, is code-based and has survived cryptanalysis attempts for over 40 years.

Multivariate cryptography: These systems typically work by trying to solve systems of multivariate polynomials over finite fields, which is known to be

Table 9.1

Classical Cryptographic Standards and Accessing the Risk from Quantum Computers

Crypto Type	Algorithm	Variants	Key Length (bits)	Classic Strength (bits)	Quantum Strength (bits)	Vulnerabilities	Recommended Quantum Resistant Solutions
Assymetric	ECC	ECC 256	256	128	0	Broken by Shor's algorithm	
		ECC 384	384	256	0		
		ECC 521	521	256	0		
	FFDHE	DHE 2048	2048	112	0	Broken by Shor's algorithm	
		DHE 3072	3072	128	0		
	RSA	RSA 1024	1024	80	0	Broken by Shor's algorithm	Merkle signature scheme Crystals-Kyber
		RSA 2048	2048	112	0		
		RSA 3072	3072	128	0		
Symettric	AES	AES 128	128	128	64	Weakened by Grover's algorithm	Larger key sizes are needed
		AES 192	192	192	96		
		AES 256	256	256	128		
	SHA2	SHA 256		128	85	Weakened by Brassard et Al's algorithm	Larger hash values are needed
		SHA 384	-	192	128		
		SHA 512	-	256	170		
	SHA3	SHA 3 256	-	128	85	Weakened by Brassard et Al's algorithm	
		SHA 3 384	-	192	128		
		SHA 3 512	-	256	170		

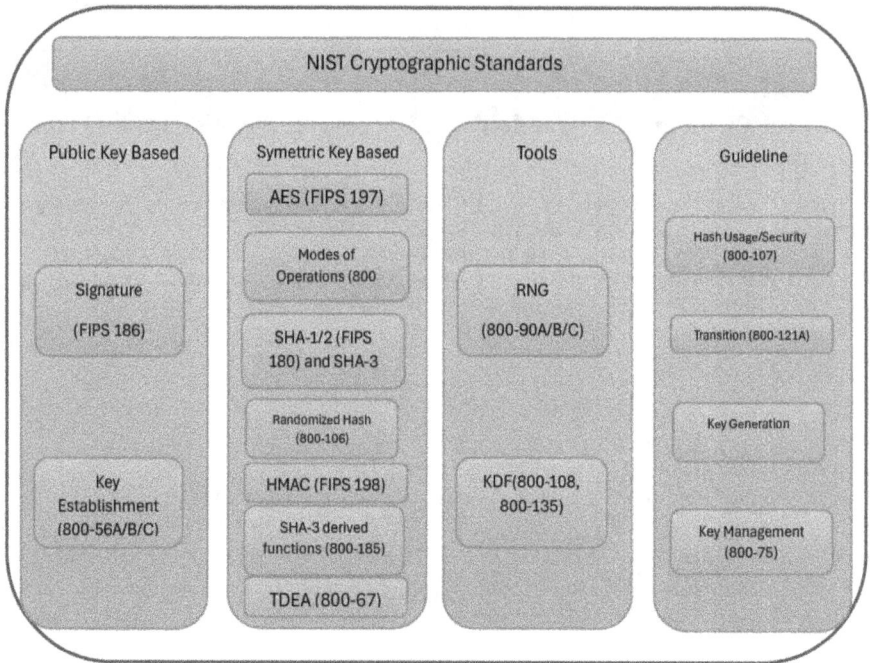

Figure 9.1   NIST cryptographic standards.

NP-hard. Few of the notable multivariate systems is the Unbalanced Oil and Vinegar (UOV) and Rainbow scheme.

Hash-Based Cryptography: Hash-based signatures are the oldest form of PQC. They are based on cryptographic hash functions, such as SHA-256, which are believed to be quantum-resistant. The Merkle signature scheme is a well-known hash-based cryptographic scheme.

Isogeny-based cryptography: This is a relatively new field, with the principal system being the Supersingular Isogeny Key Exchange (SIKE). It's based on the hardness of finding isogenies (maps) between supersingular elliptic curves.

Lattice-based cryptosystems comprise a set of dimensional points having a periodic structure. Post-quantum signature techniques guarantee proper authentication and low counterfeiting [15]. It is worth mentioning that even post-quantum algorithms may still be susceptible to other types of attacks, like side-channel and cryptanalysis attacks [3]. While side-channel attacks exploit the information leaked during the execution of a cryptographic algorithm, cryptanalysis aims to break the encryption or signature schemes by identifying structural weaknesses in the algorithm. Figure 9.2 shows these attacks.

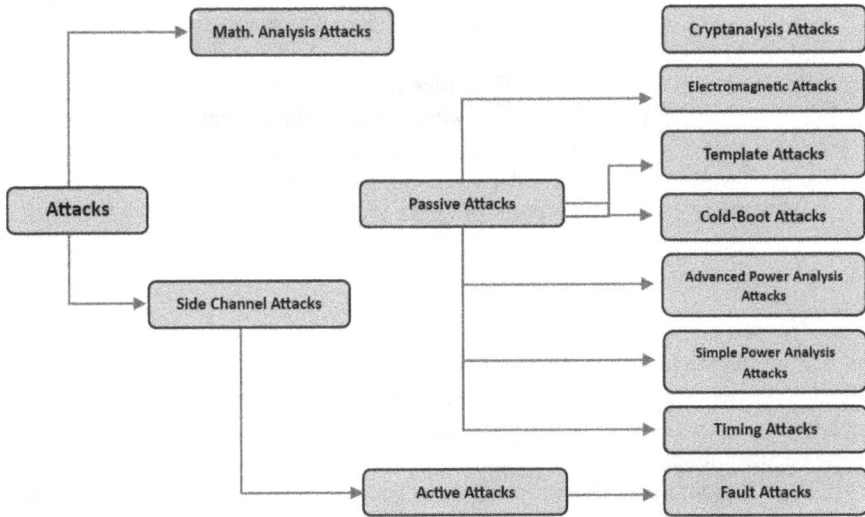

Figure 9.2 Taxonomy of attacks for Quantum Safe Cryptographic Candidates (selected in NIST's 4th round).

9.2.2 QUANTUM RANDOM NUMBER GENERATORS (QRNG)

As covered in the previous chapter, true randomness is vital for cryptography. The security of encrypted data hinges on the key's randomness – the more random the key, the harder it is to guess. Traditional computing systems, by their deterministic nature, fail to generate truly random numbers, which is where QRNG come to the rescue.

QRNG exploit quantum mechanics, specifically the inherent uncertainty principle, to generate genuinely random numbers. A common implementation involves measuring the path of individual photons – which is inherently random – to generate quantum randomness.

A range of organizations and companies are using quantum mechanics to produce random numbers, from the realm of cybersecurity to lottery draws. Swiss-based company 'ID Quantique' is one of the global leaders in the commercialization of QRNG, with its products being used in high-security contexts like key generation for encryption.

9.2.3 QKD TO THE RESCUE?

As covered in detail in previous chapters, QKD is a beacon of hope for secure communications. It offers provable security based on the laws of quantum mechanics and could resist future quantum attacks. QKD allows two parties to share a secret key that can be used to encrypt and decrypt messages. If an eavesdropper tries to intercept the quantum key during its distribution, it instantly changes its state due to the principle of quantum indeterminacy [4], and the eavesdropping can be detected.

While theory and lab-based experiments have demonstrated QKD's functionality, rolling it out in real-world applications has been challenging thus far. Limitations arise from photon loss in fiber optic cables over long distances, and the equipment required is sensitive and expensive. However, steady technological advancements are overcoming these impediments. 'ID Quantique' and 'Toshiba' are among several organizations worldwide that have deployed commercial QKD systems, with the latter recently setting a new world record for QKD over fiber optic cables.

9.2.4 PRACTICAL QKD

Securing practical quantum communication, including QKD, across real-world channels is a vital step forward for quantum-resistant cryptography. QKD is considered ideally secure when it uses perfect single photon sources and detectors. Unfortunately, ideal devices never exist in practice. As a result, device imperfections may raise security loopholes or side channels, which can break the security of practical QKD. QKD is vulnerable to various attacks such as Joint Attack, Individual Attack, Timing Attack, Collective Attack, Large Pulse Attack, Photon Number Splitting Attack, and Trojan Horse attacks [5].

The other challenge with QKD is to scale over long distances. Due to the inherent high channel loss in fibre, and decoherence, there are limitations to the distance over which secure QKD communications can occur without substantial error. This is due to the key rate of QKD which drops significantly over long distances. Various solutions, like quantum repeaters, quantum privacy amplication [6], and trusted relay schemes have been offered to solve for this problem; but none so far are foolproof.

Various protocols have been designed as solutions to secure against such device imperfections. Decoy-state QKD, Measurement-Device-Independent (MDI) QKD [7] are some examples. While the former helps prevent BB84 [8] type QKD schemes against source-based attacks like photon number splitting, the latter is a more viable solution to protect against loopholes and attacks such as time-shift attack, detector blinding attack, and dead time attack that target the detection side. This makes the MDI-QKD protocols practical and suitable for Metropolitan QKD networks than can extend upto 400 Kms [9].

The development of commercial QKD systems for long distance networks that rely on Continuous Variable (CV) QKD using better single-photon detectors with lower dark count rates [10] is a welcome advancement. Networks secured by QKD are already existent in Austria, China, and Japan. In 2020, the Tokyo-based Quantum Cybersecurity Innovation Hub announced the successful deployment of QKD to secure critical communications during the 2020 (postponed to 2021) Olympic Games.

9.3 THE QUANTUM APOCALYPSE: WHAT HAPPENS WHEN QUANTUM COMPUTERS BREAK TRADITIONAL CRYPTOGRAPHY?

The term "Quantum Apocalypse" is used to evoke the potential catastrophe that could befall when sufficiently advanced quantum computers become a reality. The threat is that they could render widely used encryption algorithms obsolete, resulting in the potential unraveling of online security systems globally.

RSA and ECC, cryptographic algorithms underpinning a significant chunk of our digital life, including banking transactions, confidential emails, and e-commerce, would be vulnerable to quantum attacks, leading to a potentially chaotic situation. However, it's essential to recognize that this "Apocalypse" is not projected to happen anytime soon. Experts estimate that it could be at least a decade, or by 2030 (conservative timeline), before quantum computers capable of breaking such cryptographic systems are developed, giving us some breathing room to prepare.

Nevertheless, the seriousness of the potential consequences of the 'Quantum Apocalypse' makes it imperative to invest in and develop quantum-resistant cryptographic systems right now – a movement captured in the urgent pace of work in the field of PQC. Much like the hysteria that preceded the transition into the new millennium, often referred to as the Y2K bug, the "Quantum Apocalypse" has been raising global concerns. This impending event describes a theoretical point in the future when quantum computers become powerful and commonplace enough to break conventional cryptographic codes. Just as the Y2K bug threatened to disrupt computer systems worldwide, the quantum apocalypse could potentially produce devastating effects across numerous sectors, causing widespread chaos in our highly digitized world.

9.3.1   COLLAPSE OF INTEGRATED ECOSYSTEM

The serenity of the interconnected world, as we know it today, is under severe threat from quantum computing advancements. Our reliance on digital technology across socio-economic spectra, from financial institutions and transportation networks to healthcare systems and remote work setups, is predicated on the security of our data.

The quantum apocalypse could unravel these interconnected systems. Examples include:

1. **Financial institutions**: Financial systems are heavily reliant on cryptography for secure transactions. A quantum computer capable of breaking today's encryption methods could undermine the integrity of financial transactions, leading to massive financial fraud and destabilizing the global financial system.
2. **Transportation networks**: Secure communications underpin the smooth functioning of our transportation infrastructure – air, road, and rail. The availability of reliable and secure navigation data, for instance, is crucial to aircraft safety. In the era of self-driving cars, the encryption of their communication systems is vital to prevent accidents. A quantum apocalypse could disrupt these essential services, leading to potentially catastrophic consequences.
3. **Critical healthcare systems**: The healthcare sector is increasingly digitized, from patient records to remote surgeries. All of this data needs to be securely transmitted and stored, requiring robust cryptography. A breakdown of digital security could compromise patient privacy or potentially even risk lives by tampering with critical healthcare tools.

4. **Remote workers**: The mass shift to remote work in the wake of the COVID-19 pandemic has highlighted the value and vulnerability of Virtual Private Networks (VPNs). These tools, which provide a secure connection to the internet, rely heavily on cryptography.

The efficient operation and coordination of basic services such as power plants, water treatment facilities, and gas pipelines rely heavily on encrypted information exchange. A disruption of this exchange could lead to service delivery failures, with severe impacts on utilities and the broader economy.

Likewise, the functionality of global supply chains, from the movement of food grains to retail products, depends on secure digital systems. Confidential business data, logistics information, and just-in-time delivery vendors all utilize encrypted communications to operate effectively and efficiently. The quantum apocalypse could severely disrupt these supply chains, leading to shortages, increased costs, and economic uncertainty.

The internet, as we know it, relies heavily on encryption for transmitting data securely. In a post-quantum world, the protocols that safeguard our online privacy and secure web traffic could be compromised, ushering in the so-called 'death' of the current internet. Ahead of this predicted cataclysm, work is already underway to build quantum-safe cryptographic algorithms that will secure the internet in the era of quantum computers with NIST releasing three PQC approved algorithms in August 2024. We are likely to witness a transformation rather than the termination of the internet, as new cryptographic solutions replace vulnerable ones, leading to the rebirth of a new, quantum-safe internet.

The listed scenarios can be avoided by pre-emptively transitioning to quantum-safe cryptographic standards, investing in quantum-resistant technologies such as QKD and fostering quantum education and awareness. By acknowledging and preparing for the quantum threat, we can mitigate its ill effects, turning a potential apocalypse into an era of new opportunities and advancements.

9.3.2   THE THREAT LANDSCAPE AND TRANSITION TO PQC

Even though the quantum threat is not imminent, the transition to post-quantum encryption requires early preparation due to the extensive infrastructural changes it entail. This is an acute concern, given the 'harvest now, decrypt later' threat model, where adversaries could capture encrypted information now for decryption as soon as quantum computers come online.

Outlined in Figure 9.3 is a map showing a research outlined in EYs Global Quantum Computing LAB Quantum Cyber readiness paper, and it provides a depiction of risk associated with current cryptographic algorithms and how they are prone to attack by quantum algorithms.

The EY report [17] suggests that even though quantum technologies aren't fully developed yet, current data can still be vulnerable to QC decryption as the life of the data may be longer than the time required to break the current encryption methods. Bad actors could simply collect a company's data today and copy it (encrypted with

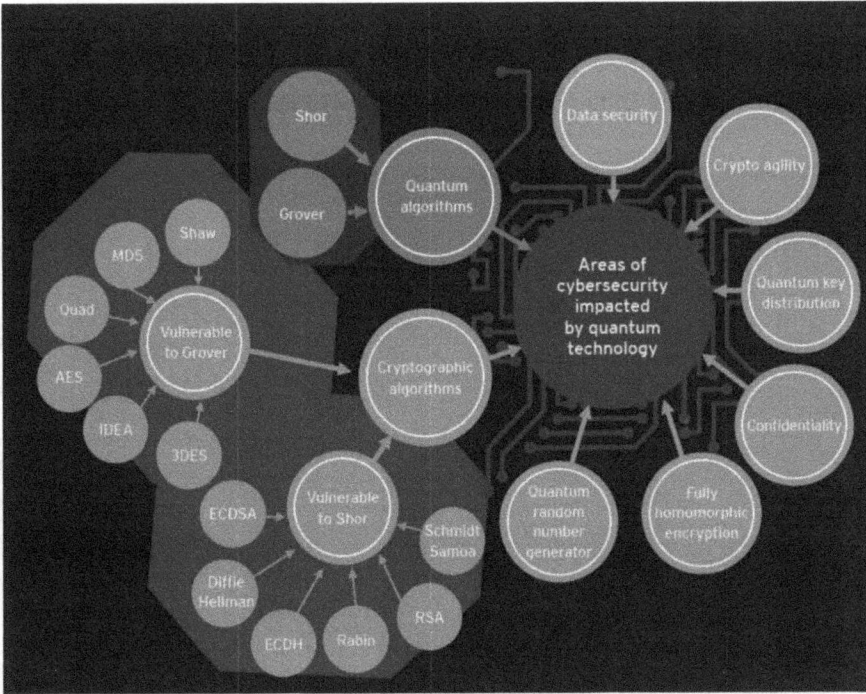

Figure 9.3 Quantum risks for cryptographic algorithms. *(From: EY Quantum Computing LAB cyber readiness paper [17].)*

current ciphers) and then decrypt it using the quantum algorithms that run on their quantum computers once they become powerful enough to crack the current cryptographic schemes. Sensitive government data that has long R&D lifecycle times and production data such as that used in automotive, aerospace, and life sciences are considered extremely vulnerable. The report goes on to outline as represented in Table 9.2, the risk associated for various cybersecurity products in use that are most likely to be impacted by quantum enabled data theft.

IBM's cost of data breach report published in 2023 [18] shows that organizations are losing as much as 60 million records in a year. The average time to detect such a data breach is between 6 to 11 months for organizations that have no attack surface management tools versus those that have. This data point is crucial because it means that though organizations who got attacked claim to have recovered from the data breach by securing systems, there is an underlying assumption that the data breached was encrypted and hence safe. But the real impact of this breach may be known only a few years later when quantum systems powerful enough could be used to possibly decode the encrypted compromised data.

To address these threats, research into PQC, cryptographic algorithms resistant to quantum attacks, is underway. PQC primarily explores mathematical problems

Table 9.2

Quantum-enabled Data Theft on Cybersecurity Products

Public Key Infrastructure	Certification Authority (CA), SSL Certificates commonly used. Since 2014, nearly all commercial CAs uses RSA public keys of at least 2048 which is considered to be breakable.
Secure Software Distribution	Mostly public key-based digital signatures, containing RSA public keys.
Federated Authorization	Single-sign-on method such as OAuth, OpenID, SAML, among others are widely based on HTTP and once hacked are extremely vulnerable to data theft and criminal acts.
Key Exchange over Public Channel	Key-sharing only between individuals Key exchange, key agreement methods are used in network security protocols like SSHE, IKE, IPsec SSL, TLS to protect private communication. Rely to a large extent on RSA, elliptic curve cryptography or Diffie-Hellman (ECDH) algorithms.
Secure Email	Secure emails commonly via S/MIME for predominantly government entities and regulated enterprises to exchange confidential/authentic email. They largely rely on RSA public keys.
Virtual Private Network	IPSec ensures company network access, work related application access, mobile workforce. VPNs can also be used to circumvent local internet restrictions in foreign countries, creating a tunneling network enabled via RSA or ECC with key establishment protocols such as IKE or mobile IKE.
Secure Web Browsing	Secure-lock web browsing via SSL/TLS enabled websites, mostly required by regulatory requirements/compliance due to user's private information, such as payment data. RSA is still the most common authentication key.
Controller devices	In-built cryptography of controller devices in any kind of machinery (cars, airplanes, manufactory, etc.) usually don't have the storage, computing or communication capabilities to support cryptographic methods such as lattice-based ones, and they are often quite difficult to replace.
Private Blockchain Transactions	Blockchain protection algorithms include RSA and ECDSA, thus the crypto world must overcome factoring problem algorithms in order to remain secure. Blockchain transaction signatures for identification and blockchain nodes with internet communication are extremely vulnerable.

Source: EY Quantum Computing LAB cyber readiness paper[17]

believed to be resistant to quantum computer attacks but efficient enough for classical computers. This includes lattice-based, code-based, multivariate-quadratic equations, supersingular elliptic curve isogeny, and hash-based cryptography which have been introduced earlier in this chapter.

The NIST is actively running a standardization process for PQC, the first three PQC algorithms published in August 2024. The standardization of PQC signifies the urgency and magnitude of the transition required to secure the digital landscape against the quantum apocalypse. NIST had selected four PQC algorithms for standardization [11]. One of these is for public-key encapsulation mechanism (KEM), while the other three are digital signature schemes:

1. CRYSTALS-KYBER: This is a key-establishment algorithm. It is designed for general encryption purposes such as creating secure websites. It is based on the hardness of the *'Module Learning With Errors'* problem, a variant of the well-studied *'Learning With Errors'* problem, which is believed to be hard even for quantum computers.

2. CRYSTALS-Dilithium: This is a digital signature algorithm. It is designed to protect the digital signatures we use when signing documents remotely. Like CRYSTALS-KYBER, it is also based on the hardness of the *'Learning With Errors'* problem.
3. FALCON: This is another digital signature algorithm. It has been standardized by NIST since there may be use cases for which CRYSTALS-Dilithium signatures are too large.
4. SPHINCS+: This is a stateless hash-based digital signature algorithm, which means that the sender does not need to store any state information between signing operations. The small public key size and fast verification size used by SPHINCS+ makes the algorithm well-suited for use in constrained environments like in embedded devices or in high-performance computing environments.

Researchers explore different parameter sets within selected NIST PQC algorithms to find a balance between signature size, computational speed, and security levels suitable for certain applications. In addition to implementing these algorithms in software, designing specialized hardware to optimize computationally heavy elements in PQC algorithms is also an ongoing area of research, which will help boost the performance of signing operations.

9.3.3 THE TRANSITION TO THE PQC ERA

Transitioning to such PQC architectures is and remains a challenge for all organizations.

To implement post-quantum security in an enterprise, a methodological approach is required. Many mechanisms are covered in previous chapters, nevertheless, a recap of some of the methodologies is provided to the reader for clarity:

1. Deploy PQC Algorithms:

    a. Replace existing public-key algorithms (RSA, ECC) with PQC algorithms that are resistant to quantum attacks [7].
    b. Consider lattice-based, code-based, or multivariate-based algorithms recommended by NIST.
    c. Integrate PQC into:
        i. VPNs
        ii. Digital signatures
        iii. Key exchange
        iv. Secure communication channels

2. Establish a Trusted Network:

    a. Create a dedicated network segment for PQC-protected communications.
    b. Utilize multi-layer encryption with a combination of algorithms to encrypt and decrypt packets.
    c. Restrict access to only authorized devices and users.

3. Implement Hardware Security Modules (HSM):

   a. Store and manage PQC keys securely within HSM.
   b. Protect keys from unauthorized access and tampering.
   c. Ensure strong physical security for HSM.

4. Enhance Key Management:

   a. Develop a robust PQC key management strategy.
   b. Address key generation, distribution, storage, rotation, and revocation.
   c. Consider using a PQC-specific key management system.

During the transition to PQC, it is worthwhile for organizations to consider a hybrid approach that combines traditional and PQC algorithms to ensure compatibility with existing systems and third-party services.

9.3.4 QUANTUM ATTACKS – HOW DO WE REMAIN PROTECTED?

As soon as practical quantum computers become available, quantum attacks on classical cryptographic algorithms will become feasible. These largely theoretical attack techniques are based on quantum algorithms such as Shor's and Grover's, details of which are described in previous chapters.

As a recap, Shor's algorithm offers a fundamentally different approach to factoring large numbers, which, if implemented on a fault-tolerant quantum computer, poses a significant threat to RSA and ECC, thereby necessitating a transition to quantum-resilient encryption methods. Similarly, Grover's quantum search algorithm, when applied in a cyber attack context, can search large databases or find the correct keys in a cryptographic system significantly faster than classical search methods.

These potential quantum attacks demonstrate the necessity of developing a sustainable quantum-safe cybersecurity strategy. As the quantum threat landscape expands, defenses against these forms of attacks are crucial, from PQC to QKD and other quantum-resistant techniques. Potential solutions may exist outside the currently prevalent lattice-based cryptography field. Areas like *isogeny-based crypto* or *code-based crypto*, which were explained briefly earlier in this chapter, are key mentions.

In the following, we break down how isogeny-based cryptography works, its mathematical foundation, and compare it to lattice-based methods.

Elliptic curves: Relies heavily on the properties of special geometric objects called elliptic curves defined over finite fields. Elliptic curves look to address foundational problems that concern finding short vectors within high-dimensional lattice structures. These mathematical entities involve linear algebra concepts.

Isogenies: These are structure-preserving maps between elliptic curves. Imagine them as bridges that connect different, yet mathematically related, elliptic curves. The mathematics behind isogenies is rooted in algebraic geometry and the study of elliptic curves. Security assumptions focus on difficulties regarding isogeny pathfinding problems.

The Hard Problem: Given two elliptic curves, computing the specific isogeny (the exact bridge) that relates them is computationally very difficult. This forms the basis for the security of these systems.

One example of where Isogeny-based cryptographic solutions can be implemented are to secure "Chip-based transactions". This is mainly due to the small key, and signature size of these schemes. A scheme worth referencing is the "Supersingular Isogeny Diffie-Hellman (SIDH)" key exchange that exhibits relatively compact key and signature sizes. SIDH relies on constructing isogenies between supersingular elliptic curves defined over a finite field with a special prime characteristic. This aligns well with memory constraint requirements of chip-based secure designs.

Another isogeny-based signature proposal scheme that is quite promising is the CSI-FiSh scheme. CSI-FiSh is also based on isogenies, similar to SIDH. However, it also employs techniques from code-based cryptography (Niederreiter scheme) and uses public matrices representing a syndrome decoding problem of an error-correcting code.

A key issue in the current data encryption paradigm is that data has to be decrypted prior to performing mathematical computation or using it for data processing purposes. The next generation of encryption technologies is trying to fix this problem by developing encryption techniques that enable data computation on encrypted data. This paradigm will also help the post-quantum computing data security scenarios. Currently, there are two primary next-generation encryption techniques, Fully Homomorphic Encryption (FHE) and Multi-Party Computation (MPC). These techniques have delivered promising results in performing data computation while the data is still encrypted [12].

Quantum cryptanalysis may still be a decade away, but some secrets might retain their value for many decades. Given the longevity of some secrets, there is no room for complacency about the quantum threat. While quantum-safe security via QKD is bound to be a motivation for investing in quantum networking, Quantum networks may enable some applications that are simply infeasible with classical networks. These include encryption schemes allowing users to certify the deletion or retention of data, detect tampering, and create unique time windows for decryption [13].

Remaining protected from quantum attacks requires embracing new encryption approaches, refining and deploying quantum-resistant cryptographic measures, and continuous research in this field to match pace with quantum computing evolution.

9.3.5 THE QUANTUM CYBER ARMY: INDUSTRY EFFORTS

Reflecting on the threats posed by quantum computing, various industry sectors are gearing up to prepare for potential quantum attacks. Companies such as Microsoft, IBM, Google, and others are not just at the frontier of quantum computing research, they're also working toward developing quantum-resistant algorithms.

In addition to Microsoft's development of Quantum Development Kit, which supports Q# language, Microsoft Research is also actively working in the field of PQC and has proposed SIKE for consideration in NIST's PQC standardization process. Similarly, IBM is an active contributor to the Open Quantum Safe (OQS) project, which aims to support the development and prototyping of quantum-resistant cryptography. Google has not been far behind, having recently implemented the post-quantum algorithm, 'New Hope', in the Chrome browser for a small number of connections between desktop Chrome and Google's servers as a real-world test of its practicality.

Government organizations worldwide are also actively investing in quantum research and PQC. NIST is currently in the process of selecting algorithms that could help usher in the quantum-proof era. The OQS project has been developing and prototyping quantum-resistant cryptography and integrating it in an open-source library appointed to all computational scientists and security practitioners. The efforts of European Union Agency for Network and Information Security (ENISA) are focused on protocols which can interoperate with existing communication networks and are secure against both quantum and classical attacks (Di Franco). The quantum-safe cryptography project by European Telecommunications Standards Institute (ETSI) is directed to the security of government and military communications, financial and banking transactions, the storage of personal and confidential corporate data in the cloud [14].

Concurrently, government bodies in the EU and China are similarly investing heavily in quantum research. Some notable mentions are the 10-year 'Quantum Flagship program', which was announced in May' 2016 and launched in 2018 by the European Commission who has planned to invest one billion Euros for over 10 years to accelerate quantum technologies' development, China's quest to succeed in secure QKD in the quantum satellite networks, 'Quantum Canada' initiative announced as part of the Government of Canada's National Quantum Strategy in January 2023 to facilitate collaboration and coordination in quantum R&D and PQC standards, and Australia's Centre for Quantum Computation and Communication Technology that is focused on developing novel quantum computation system.

In essence, a "Quantum Cyber Army" is being formed across sectors globally, with a mutual interest in harnessing quantum computing's benefits and mitigating its threats. As quantum computing continues to evolve, it's crucial that this cross-sector collaboration continues, addressing the challenges and ensuring cybersecurity in a post-quantum world.

9.4 CONCLUSION

In conclusion, the concept of "Harvest Now, Decrypt Later" underscores the critical urgency in addressing quantum cybersecurity threats. As the quantum horizon approaches, it is imperative that a forward-looking cybersecurity posture be adopted to mitigate the risks associated with quantum computing's potential to undermine current cryptographic standards. Initiatives to develop quantum-resistant encryption must be fast-tracked and widely implemented to safeguard data against future decryption threats. Collaborative efforts between academia, industry, and government agencies are crucial to establish robust quantum-secure frameworks that ensure the continued confidentiality, integrity, and availability of sensitive information. By anticipating the challenges and acting decisively, we can proactively secure our digital infrastructure against the looming quantum threat, thereby preempting the dangers of adversaries capitalizing on encrypted data harvested today.

REFERENCES

1. Yin, Juan, Li, Yu-Huai, Liao, Sheng-Kai, Yang, Meng, Cao, Yuan, Zhang, Liang, Ren, Ji-Gang, Cai, Wen-Qi, Liu, Wei-Yue, Li, Shuang-Lin, Shu, Rong, Huang, Yong-Mei, Deng, Lei, Li, Li, Zhang, Qiang, Liu, Nai-Le, Chen, Yu-Ao, Lu, Chao-Yang, Wang, Xiang-Bin, Xu, Feihu, Wang, Jian-Yu, Peng, Cheng-Zhi, Ekert, Artur K., & Pan, Jian-Wei. (2020). Entanglement-based secure quantum cryptography over 1,120 kilometres. Nature, 582 (7813), 501–505. https://doi.org/10.1038/s41586-020-2401-y

2. Bergou, J. A., Hillery, M., & Saffman, M. (2021). Decoherence and quantum error correction. https://doi.org/10.1007/978-3-030-75436-5_9

3. Baseri, Y., Chouhan, V., & Ghorbani, A. (2024). Cybersecurity in the quantum era: Assessing the impact of quantum computing on infrastructure. arXiv preprint arXiv:2404.10659.

4. Fierke, K. M., & Mackay, N. (2020). To 'see' is to break an entanglement: Quantum measurement, trauma and security. Security Dialogue, 51(5). https://doi.org/10.1177/0967010620901909

5. Sharma, N., & Ketti Ramachandran, R. (2021). The emerging trends of quantum computing towards data security and key management. Archives of Computational Methods in Engineering, 28 (7). https://doi.org/10.1007/s11831-021-09578-7

6. Deutsch, D., Ekert, A., Jozsa, R., Macchiavello, C., Popescu, S., & Sanpera, A. (1996). Quantum privacy amplification and the security of quantum cryptography over noisy channels. Physical Review Letters, 77(13). https://doi.org/10.1103/PhysRevLett.77.2818

7. Lindsay, J. R. (2020). Demystifying the quantum threat: infrastructure, institutions, and intelligence advantage. security studies, 29(2). https://doi.org/10.1080/09636412.2020.1722853

8. Yevseiev, S., Gavrilova, A., Tomashevsky, B., & Samadov, F. (2019). Research of crypto-code designs construction for using in post quantum cryptography. Development Management, 16(4). https://doi.org/10.21511/dm.4(4).2018.03

9. Zhang, Q., Xu, F., Chen, Y.-A., Peng, C.-Z., & Pan, J.-W. (2018). Large scale quantum key distribution: Challenges and solutions [Invited]. Optics Express, 26(18). https://doi.org/10.1364/oe.26.024260

10. Diamanti, E. (2015). Practical secure quantum communications. Quantum Optics and Quantum Information Transfer and Processing, 2015, 9505. https://doi.org/10.1117/12.2185062

11. Sokol, S. (2023). Navigating the quantum threat landscape: Addressing classical cybersecurity challenges. Journal of Quantum Information Science, 13(2). https://doi.org/10.4236/jqis.2023.132005

12. Badhwar, R. (2021). The CISO's Next Frontier: AI, Post-Quantum Cryptography and Advanced Security Paradigms. https://doi.org/10.1007/978-3-030-75354-2

13. Lindsay, J. R. (2020). Surviving the quantum cryptocalypse. Strategic Studies Quarterly, 14(2). https://www.airuniversity.af.edu/Portals/10/SSQ/documents/Volume-14_Issue-2/Lindsay.pdf

14. Jelčić Dubček, D. (2021). Quantum computers – An emerging cybersecurity threat. Annals of Disaster Risk Sciences, 3(2). https://doi.org/10.51381/adrs.v3i2.56

15. Serrano, M. A., Sanchez, L. E., Santos-Olmo, A., Garcia-Rosado, D., Blanco, C., & Fernandez-Medina, E. (2023). Towards a quantum world in cybersecurity land. CEUR Workshop Proceedings, 3408.

16. Faruk, M. J. H., Tahora, S., Tasnim, M., Shahriar, H., & Sakib, N. (2022). A review of quantum cybersecurity: Threats, risks and opportunities. 2022 1st International Conference on AI in Cybersecurity, ICAIC 2022. https://doi.org/10.1109/ICAIC53980.2022.9896970

17. Wong, J., & Kristin, G. (2023). EY Quantum Computing LAB cyber readiness paper. https://www.ey.com/content/dam/ey-unified-site/ey-com/en-gl/insights/innovation/documents/ey-quantum-approach-to-cybersecurity-v3.pdf

18. IBM Cost of Data Breach Report 2023 https://www.ibm.com/reports/data-breach

19. Di Franco, F. (2018). Analysis of the European R&D priorities in cybersecurity. European Union Agency for Network and Information Security (ENISA). ISBN: 978-92-9204-278-3. https://www.enisa.europa.eu/publications/analysis-of-the-european-r-d-priorities-in-cybersecurity

10 Post-Quantum Cryptography in Embedded Systems, Future Trends, and Research Directions

Amer Mosally, Ibrahim Kabir,
and Mohammad Hammoudeh

10.1 INTRODUCTION

Quantum computing is a relatively new computing approach that uses the laws of quantum mechanics to develop and design computer-based technology. Information theory, computer science, mathematics, and physics principles explain energy and matter behavior at atomic and subatomic levels. Unlike conventional computers, quantum computing offers higher computing capacity, lower energy utilization, and high speed [1]. It has the potential to revolutionize computing power and has applications in various fields. It is important to note that quantum computing does not seek to replace classical computing but aims to augment it by enabling significant acceleration in certain applications, surpassing the capabilities of classical computing. Although a fully operational quantum computer has yet to be realized, numerous ongoing initiatives and projects are contending to attain this ambitious objective [2, 3].

An embedded system is often defined as a microprocessor-based computer hardware system with software and peripheral devices with a dedicated function within a larger mechanical or electronic system. It has distinguishing features of being application-specific, doing real-time operations, being resource-constrained, and being able to be integrated into larger systems. Embedded systems are vital because they are components of virtually all systems nowadays due to their cross-functionality and ability to enhance them [4].

Given the capabilities of quantum computers, the security of embedded systems is at significant risk [5, 6]. Quantum computers would make it much easier to access protected information by breaking encryption, which would otherwise have been resistant to attacks by traditional computers. This is more evident and pronounced in embedded devices with limited processing power and memory to hold and process the best encryption algorithms. It is necessary to develop quantum-safe technology in anticipation of the actualization of quantum computing in the near future.

This chapter offers an introduction and overview of quantum cryptography as it pertains to embedded systems, highlighting the critical importance of quantum-safe cryptography in the post-quantum era. It reviews existing literature, explains pertinent theories, and presents key research findings, thereby identifying gaps in the current body of research on quantum-safe embedded systems. Through case studies and illustrative examples, the chapter explores both current and potential future applications of quantum-safe embedded systems. It concludes by highlighting the key challenges and emerging trends in this field, providing insightful recommendations and suggesting promising directions for future research.

10.2   IMPACT OF QUANTUM COMPUTING ON CRYPTOGRAPHY

In a classical computer system, all the data is stored and represented using bits. A bit is the smallest measure of data in basic computer theory, and the value of the bit is either 0 or 1 at a time. On the other hand, quantum computing data is represented by a quantum bit, the smallest unit called a qubit. Qubit state is defined inside the Ket symbol "$|0\rangle$ or $|1\rangle$". Quantum computers apply quantum physics principles to perform computations, differentiating them from classical computers.

The fundamental principle of quantum mechanics is quantum entanglement. Quantum entanglement can be described as a phenomenon where two particles are connected in a specific way and both exist in superposition states. A superposition state in quantum mechanics exists in multiple states simultaneously, and any changes to the quantum particle state will instantly reflect in the state of the other particle even if they are separated. All the computed answers from a quantum computer are embedded in superposition states. This quantum principle state refers to the ability of a qubit to be in a state that is a mixture of the '0' and '1' states simultaneously. Only one value is given when the answer is measured, and the other information will be lost. To illustrate, the qubit will be in the superposition states of $|0\rangle$ and $|1\rangle$, but measuring the state will collapse the superposition state to a single state, either 0 or 1.

Public key cryptography relies on complex math problems such as RSA, which depends on prime factorization, and Diffie-Hellman, which uses discrete log problems. A quantum computer can quickly solve these problems and break down the associated cryptography. We need to use complex math problems to build new quantum-resistant cryptography, which is hard even for quantum computers.

The current public-key cryptography systems are vulnerable to quantum computing cryptanalysis using Shor's algorithm. The algorithm is designed to solve specific mathematical problems much more efficiently than classical computers. Shor's algorithm leverages the quantum bits to operate and has the potential to break the current level of cryptography. To illustrate, modern classical encryption, such as RSA, relies on the hardness of factorizing large prime numbers, which is the public key. The public key is just a multiplication of two large prime numbers: the private key of the sender and the receiver. The algorithm can break RSA by utilizing quantum bits (qubits), representing a superposition of states. This allows the algorithm to perform many calculations simultaneously. Grover's algorithm provides a quadratic speedup for unstructured search problems compared to classical algorithms, and quantum computers can use it to speed the brute force attacks on symmetric ciphers.

10.3   POST-QUANTUM CRYPTOGRAPHY STANDARDIZATION INITIATIVES

Initiatives to provide a standard framework for implementing Post-Quantum Cryptography (PQC) have been developed by many organizations. Some of them are the IETF initiative, which involves the Internet Engineering Task Force (IETF). Another organization is the European Telecommunications Standards Institute (ETSI) Quantum-Safe Cryptography (QSC) working group, which aims to recommend and assess PQC algorithms and consider implementation challenges [7]. Other such ones are proposed by the Institute of Electrical and Electronics Engineers (IEEE) through the P1363 project, which produced the IEEE 1363.1-2008 standard for lattice-based public-key cryptography [8], and by the International Organization for Standardization (ISO), which makes standards for cybersecurity and is developing PQC through the ISO/IEC JTC 1/SC 27 project [9]. The American National Standards Institute (ANSI) has also published a white paper on the security hazards of quantum computing for the banking industry [10].

The most widely accepted and advanced standardization initiative is that by the National Institute of Standards and Technology (NIST), an agency in the United States. The NIST is evaluating proposals for PQC to address the quantum threat [11].

In 2017, NIST announced a call for PQC algorithm submissions to be selected as new PQC standardization after a multi-step process. The process of selection post-quantum methods involved four rounds of competition and review, and 15 algorithms qualified for the third round. NIST announced the final list of their standards in 2022, which are CRYSTALS-Kyber, CRYSTALS-Dilithium, FALCON, and SPHINCS+. And in August 2024 three approved PQC algorithms were released by NIST. However, some PQC candidates have recently been found vulnerable to classical attacks such as the Rainbow system and SIKE, which highlights the existing gaps in the PQC and a need for more research in this area [12].

10.4   TYPES OF PQC ALGORITHMS

All data transmitted over the internet is technically accessible to the public, but it is protected through encryption algorithms like RSA. RSA is a public-key cryptosystem that secures data using the product of two large prime numbers, which are computationally infeasible to factor with current technology. However, the "harvest now, decrypt later" strategy poses a potential threat. This approach involves collecting encrypted data today with the intention of decrypting it in the future when quantum computers become powerful enough to break encryption algorithms like RSA, undermining the security they currently provide.

There are different types of post-quantum cryptographic systems (see Table 10.1). One of the most popular is lattice-based cryptosystems. It includes an

Table 10.1

Overview of Post-quantum Algorithms on NIST Third Round [13]

Public Key Encryption	
Name	**Type**
BIKE	Code-based
Classic McEliece	Code-based
CRYSTALS-KYBER	Lattice-based
FrodoKEM	Lattice-based
HQC	Lattice-based
NTRU	Lattice-based
NTRU Prime	Lattice-based
SABER	Lattice-based
Digital Signature Algorithms	
Name	**Type**
CRYSTALS-Dilithium	Lattice-based
FALCON	Lattice-based
GeMSS	Multivariate polynomial
Picnic	Symmetric-key cryptographic primitives
Rainbow	Multivariate polynomial
SPHINCS+	Hash-based
MQDSS	Multivariate polynomial

infinite set of points in a grid involving a lattice problem such as the Closest Vector Problem (CVP), which is hard to solve even with a quantum computer [14]. Code-based cryptosystems are based on error-correction codes, which rely on the difficulty of decoding a general linear code. Multivariate polynomial cryptosystems are based on the difficulty of solving multivariate equations of systems [15]. Hash-based cryptosystems involve cryptographic constructions based on the security of hash functions, and they are mainly used for digital signatures [16].

Methods that rely on CVP, like lattice-based cryptographic algorithms, use the difficulty of solving CVP to secure digital communications. CVP involves finding the closest lattice point to a target point (which may not be on the lattice) given a set of basis vectors [17]. The main hardness of CVP is in high-dimensional spaces where the difficulty of solving CVP scales exponentially with the number of dimensions. The Shortest Vector Problem (SVP) is a computationally challenging problem in lattice theory that involves finding the shortest non-zero vector in a lattice [18]. Also, the complexity of SVP increases significantly in higher dimensions.

In 2022, NIST announced the first four PQC algorithms which will eventually become part of NIST standards. They have been extensively studied and proven to be resilient post-quantum cryptographic algorithms. Three of the algorithms are for digital signatures, which are CRYSTALS-Dilithium, FALCON, and SPHINCS+, and one algorithm for public key encryption and decryption which is CRYSTALS-KYBER.

CRYSTALS-KYBER is a lattice-based and secure Key Encapsulation Mechanism (KEM) based on solving the Module Learning With Errors (MLWE) problem [19]. MLWE is a mathematical problem used in cryptography to create a secure encryption algorithm by representing the information as a set of equations with random errors. It establishes a symmetric key for protocols like TLS. The Kyber algorithm has three levels of security Kyber 512, 768, and 1024. Each level has a different key size and they are based on the SVP. Quantum computers struggle to find the shortest vector in a high-dimensional lattice to break the algorithm.

CRYSTALS-Dilithium is a digital signature scheme that is one of the numerous derivatives of the Cryptographic Suite for Algebraic Lattices (CRYSTALS) [20]. The hardness of such schemes is based on MLWE and Module Short Integer Solution (MSIS) problems. The Dilithium DSA has three different security levels, which differ in key and signature sizes (Dilithium2, Dilithium3, and Dilithium5). This DSA allows an individual to sign a message or document so that others can verify the sender's authenticity without being able to forge it using math problems to remain secure.

Fast-Fourier lattice-based compact signatures over the Nth-degree Truncated Polynomial Ring Units (NTRU) Falcon derivative is an algorithm for digital signatures based on lattice cryptography. The algorithm relies on challenging the short integer solution (SIS) problem over the NTRU [21]. Its two types are Falcon 512 and Falcon 1024 with the numbers in the name representing "n," which is a parameter that affects the size and security level of the cryptographic keys and signatures.

SPHINCS+ is a sophisticated post-quantum cryptographic algorithm that employs Winternitz one-time signature+ (WOTS+), which relies on cryptographic hash functions that turn any input into a fixed-size string of characters [22]. The private key is a collection of random numbers, and the corresponding public key is generated by applying a hash function a certain number of times to each number in the private key [23].

10.5 POST-QUANTUM CRYPTOGRAPHY IN EMBEDDED SYSTEMS

10.5.1 THREAT OF QC TO EMBEDDED SYSTEMS

The security of embedded devices depends upon the digital signature schemes which they run on. Breakthroughs in Quantum computers that can execute Shor's and Grover's algorithms will make traditional digital signature schemes no longer secure [24]. Shor's algorithm solves the mathematical problems that secure public-key algorithms, while Grover's algorithm reduces by half, the strength of Advanced Encryption Standard (AES), and Data Encryption Standard (DES) for any key length [25].

This is a serious concern for embedded computers deployed in critical infrastructure and sensitive applications particularly those with long lifetimes.

There is, therefore, an obvious need to move from running conventional cryptography on embedded systems to running PQC algorithms. However, there are challenges associated with this. The major restrictions on embedded devices are the available RAM, available ROM/Flash Storage, and Performance (clock speed) [26].

A key consideration of implementing PQC algorithms on resource-constrained embedded devices is the computational complexity of these algorithms and their bandwidth requirement. It is also important to know if the embedded devices can meet the computational requirements of these algorithms or if there is a need for dedicated hardware accelerators [27].

10.5.2 PQC PROJECTS

A lot of funding and financial support has been granted toward developing devices and technology that are resistant to Quantum attacks. These efforts have come from governments and agencies alike. A popular one is a grant by both the European Union (EU) and the Japanese Science and Technology Agency given to endeavors relating to PQC systems. These projects include, including PQCrypto, SAFEcrypto, PROMETHEUS, the Quantum RISC, and the PQM4 project.

The PQCrypto project received approximately €4 million in funding between 2015 and early 2018 from the European Union's Horizon 2020 program [28]. This initiative, conducted by researchers from the Eindhoven University of Technology in the Netherlands, entailed a collaboration with participants from many European and Asian countries. The project had three segments, with each one examining PQC for communications over the internet, cloud computing, and embedded devices, respectively.

The primary focus of SAFECrypto was on the utilization of lattice problems as a means to attain computational hardness. Similar to PQCrypto, this project was also funded by the EU, with a total funding of approximately €3.2 million. The duration of the project was from January 2015 to December 2018. The lead role in overseeing SAFECrypto was taken by the Queen's University of Belfast, United Kingdom [29]. The partnership involved collaboration with researchers and professionals from Switzerland, Germany, the United Kingdom, and Ireland. The project's software library, which shows the implementation of various cryptographic schemes, such as Ring Learning With Errors (Ring-LWE), Bimodal Lattice Signature Scheme B (BLISS-B), and Kyber, can be accessed on GitHub. The PROMETHEUS initiative commenced in January 2018 and concluded in December 2021. The task of overseeing the project was carried out by the École Normale Supérieure de Lyon, which secured a significant funding grant of approximately 5.5 million euros from the EU [30]. Collaborating with various notable public institutions and enterprises in Israel, France, Spain, Germany, the United Kingdom, and the Netherlands, the project has already disseminated a substantial collection of pertinent literature.

The QuantumRISC project involves PQC specifically for embedded devices. The main goal of the project is to optimize PQC algorithms for low-resource (memory, storage, and power) devices while trying to provide an acceptable level of security [31].

The PQM4 (PQC on ARM Cortex-M4) project is an optimization effort specifically focused on ARMv7-based devices, particularly those utilizing the ARM Cortex-M4 processor. This project aims to provide efficient and secure implementations of post-quantum cryptographic algorithms for resource-constrained IoT devices. A lot of research into implementing and optimizing PQC algorithms, including the PQM4 project, has been carried out on an ARMv7-based microcontroller board because of NIST's choice of the ARM-Cortex-M4 as one of the devices for conducting performance evaluations in the PQC competition [32]. With the rapid development of quantum computing technology, the need for quantum-resistant cryptographic algorithms has become increasingly important. This makes optimization studies essential to ensure that these algorithms can be efficiently executed on devices with ARMv7 and ARMv8 architectures [33].

10.5.3 CASE STUDIES AND PQC IMPLEMENTATIONS IN EMBEDDED DEVICES

Recently, researchers have given the implementation of PQC in embedded devices a lot of attention. There have been many attempts to implement the different NIST Key Encapsulation Mechanisms (KEMs) and Digital Signatures in a diverse range of embedded devices. Maximilian Schöffel et al. [34] implemented PQC on an IoT edge device. The latency and energy consumption of TLS Handshakes with post-quantum KEMs and Digital Signature algorithms were compared. The Latest NIST Round 3 KEMs (KYBER, NTRU, and SABER) were implemented on the client and server sides. On the client side, the algorithms were implemented using an ARM standard library for embedded devices (mbedTLS). While on the server side, the OQS-OpenSSL was updated. Analysis showed that the three candidates plus the alternate candidate, NTRU Prime, performed much better than other KEMs and required 25% more time than the conventional ECDHE. The alternate candidate SIKE's complicated calculations, especially at higher security levels, demand a lot more energy than the other approaches but perform well with hardware accelerators. The least amount of memory is taken up by SIKE_P434 (4KB) and KYBER512 (2.5KB), which makes them ideal for devices with memory constraints. The code-based KEMs, on the other hand, have the biggest memory footprints. Because of its minimal stack utilization, KYBER512 worked best in their configuration and was selected for analysis of DSAs.

The authors in [34] also conclude that the use of post-quantum DSAs in homogenous PKIs as recommended by NIST leads to a significant rise in communication delay and energy utilization. This is because although Dilithium has efficiency in making computations, it also has high bandwidth needs, and Falcon has a signing mechanism that requires a lot of processing power. Nonetheless, a specialized hardware accelerator for the client-side Falcon signature process can be implemented or

Dilithium-based client certificates together with Falcon-based server and CA certificates can be used to create an effective PKI. Battery runtime was determined not to be significantly impacted when moving from conventional cryptography to PQC.

Lukas Malina et al. [35] utilized a single ARM board with a 32-bit CPU ARMv7l, 1 GB RAM with a Linux OS (Raspbian Stretch Lite) to execute six post-quantum cryptographic KEMs which were: New Hope, NTRU, BCNS, Frodo, McBits, SIDH. Among the six, the New Hope scheme was the most efficient. The New Hope, NTRU, BCNS schemes required less than 35 milliseconds, while the SIDH scheme necessitated approximately 4414 milliseconds. The New Hope and NTRU schemes offer a trade-off between efficiency and message size parameters, and they also exchange just under 4 kilobytes during the protocol. The implementation results reveal that the runtime of New Hope was 2.36 milliseconds and was marginally less efficient than ECDH-p256, which took 2.11 milliseconds. Both implementations were done with the C programming language. The ECDH scheme written in JAVA took more time than its C counterpart, indicating that C is superior for the implementation of these schemes. The findings show the superiority of PQC schemes like lattice-based KEMs over traditional asymmetric cryptosystems for key establishment.

Rameez Asif [36] compared the memory usage (in bytes), computational time (in milliseconds), and clock cycle counts for the practical hardware implementation of Lattice-Based Cryptography (LBC) on an ARM CORTEX-M platform. Both SABER and ThreeBears variants exhibit lower bandwidth consumption at various NIST security levels, making them suitable choices for lightweight implementation of LBC in IoT networks. Four schemes were selected to implement hardware for LBC algorithms based on KEMs: Saber-5, Kyber-5, NewHopeCCA-5, and FrodoKEM-AES-3. Among these schemes, Saber-5 demonstrated the least number of cycles and the shortest time for key generation, encryption, and decryption operations. Furthermore, Saber-5 exhibited the most efficient performance in terms of cycles and time for these operations. For hardware implementation of LBC schemes based on signatures, four different LBC schemes were chosen: Falcon-1, Falcon-5, Dilithium-3, and qTesla-3. Dilithium-3 produced lower values for the mentioned parameters, indicating superior performance. The author concludes that Dilithium can be utilized in post-quantum IoT networks where achieving level-5 security is not necessarily the primary objective, but rather an acceptable range of security between levels 1 and 3 is acceptable.

Gandeva Bayu et al. [37] implemented PQC on a low-resource, Raspberry Pi device, using the Message Queuing Telemetry Transport (MQTT) protocol for data transmission. The average CPU and RAM consumption values of the RPi-4 while running the RSA, NTRU, and SABER algorithms for encryption and decryption were obtained. The best average values for both RAM and CPU consumption for encryption were 18% given by NTRU-401 and 24% given by NTRU-539 at 128 and 196-bit security levels, respectively. The best average RAM consumption for the decryption process was 3%, given by both NTRU-401 and NTRU-539 at 128 and 196-bit security levels, respectively. Also, the best average CPU consumption for the decryption process was 16% given by Light SABER and 19% given by NTRU-539 at 128 and 196-bit security levels, respectively.

In terms of CPU and memory usage, the modified implementation of NTRU outperformed SABER and RSA. On the other hand, Light SABER was the best if considering encryption and decryption delays. Notably, at the security level of 128 bits Light SABER was almost 45 times better than RSA-4028, while NTRU-401 was about five times faster. Similarly, at the 196-bit security level, SABER had a speed improvement of 8.123 times compared to the RSA-7680, and the NTRU-539 was close to 10 times faster than the RSA-7680. Therefore, it can be concluded that the employed PQC techniques on the Raspberry Pi device demonstrated efficient utilization of CPU and memory resources. The NTRU algorithm, in particular, exhibited superior performance compared to SABER and RSA, while Light SABER showcased better encryption and decryption speeds.

Kevin Bürstinghaus et al. [38] Evaluated some PQC primitives with an adapted embed Transport Layer Security library on four embedded platforms; Raspberry Pi3, ESP32, Industrial Field Option Card, and a Low Pin Count (LPC) bus device. SPHINCS+ signature with SHA-256 performed slower on Raspberry Pi3 and ESP32 than the ECDSA. Over the range of platforms, the SHAKE-256 variant took 2s on ESP32, 17.3s on FOC, and 44s on LPC, which is much slower than ECDSA. SPHINCS+ using SHA-256 needed 12,000 and 280,000 calls for verification and signing, respectively, while SHAKE-256 required 11,500 calls for verification and 260,000 calls for signing. When using the SHA-256 version of SPHINCS+ on the Raspberry Pi, signature performance was much faster (less than 1 s) than with ECDSA (15 ms). On the other hand, the SHAKE-256 version took more than 5 s. The Raspberry Pi's SPHINCS+ verification process is effective; it took 66 ms for SHA-256 and 239 ms for SHAKE-256. Unlike SPHINCS+ signatures, Kyber512 does not deplete the target platforms' computing capacity. For Kyber512, key creation, encryption, and decryption were completed in less than a millisecond. The ESP32 times are longer (12 ms for key generation, 16 ms for encryption, and 18 ms for decryption). Kyber consistently performed better than the traditional Elliptic-curve Diffie-Hellman (ECDHE) key exchange on all devices.

The runtime measurements demonstrated the efficiency of Kyber-SPHINCS+ on the Raspberry Pi for the whole TLS handshake, where the SHAKE-256 version was noticeably faster and the SHA-256 version was only slightly slower than the conventional handshake. Additionally, the memory footprint analysis indicates that Kyber-SPHINCS+ has some reasonable overhead.

10.6   FUTURE TRENDS AND RECOMMENDATION

10.6.1   RECOMMENDATIONS

After a survey of PQC in embedded systems, discussion of implementation results, and analysis of implementation challenges, it is pertinent to propose recommendations to solve underlying problems and suggest future research directions. Optimization of the performance of PQC algorithms should be device-centric. It must take into account the processing power and memory limitations of the device, even after selecting from the best PQC algorithms.

A thorough examination of the impacts of varying algorithm parameters is also necessary. Achieving the appropriate balance among critical parameters like key length, performance, speed, number of cycles for key generation, and number of bits (for pre-quantum) is instrumental in the successful implementation of post-quantum cryptographic algorithms. This requires a well-researched and experimented approach to parameter adjustment to meet the specific requirements of different embedded systems and achieve the best balance between security and performance.

Likewise, it is necessary to optimize computational resources. Hence, the development of lightweight cryptographic algorithms suitable for resource-constrained environments, such as embedded systems, is crucial. This optimization ensures that security is not compromised while accommodating the limitations of these devices.

Further research on specialized hardware designed for PQC is recommended. By doing this, signing computations will be accelerated, greatly increasing the effectiveness of PQC techniques. This may enhance embedded systems' general responsiveness and speed. Moreover, collaboration in standardization efforts is essential for the creation and acceptance of global guidelines for PQC. A unified approach through collaboration between industry, academia, and standardization agencies will promote more reliable and safe PQC implementations.

To keep up with developments in quantum computing and cryptography research, it is also essential to regularly examine and update cryptographic practices. Lastly, it is advised to develop collaborations between public and private organizations to promote creativity and ease the sharing of resources in PQC.

10.6.2 IMPLEMENTATION CHALLENGES

Large key sizes: PQCs require larger key sizes, which can be difficult to implement on embedded devices because most conventional encryption systems employ relatively small key sizes (between 128 and 4096 bits) [39]. Performance must thus be maximized by offering a trade-off between security, key length, and performance.

Slow key generation: Some post-quantum cryptosystems need the generation of new keys for each set of signed messages to prevent attacks. This puts a limit on how many messages a key can be used to sign. Because of this, key generation operations demand more computing power than those used in pre-quantum encryption [40]. For this reason, it is necessary to come up with strategies for modifying post-quantum key generation processes to save energy consumption.

Absence of standardized security level benchmarks: One persistent issue that still exists is the lack of standardized criteria for security levels. It is currently undefined what constitutes a standard definition or metric for security against quantum attacks that is agreed upon by everyone. Therefore, to attain an acceptable level of security, an agreement must be made on how to evaluate security against quantum attacks and choose appropriate key lengths by both industry and academia.

Physical Security of devices: It is essential to assess the physical security of the devices running these developed post-quantum cryptosystems to verify overall security. It is worth noting that although the suggested PQCs need to be resistant to mathematical attacks, such as those carried out by quantum computers, the algorithms' actual implementation might also be vulnerable to physical attacks. This vulnerability results from the possibility of an attacker physically accessing the embedded devices running these algorithms [41]. Therefore, the prevention of physical attacks must be given high priority during the design and assessment of the suggested cryptosystems.

Ongoing standardization: Because of ongoing efforts of standardization, and improvement of PQCs that may continue well into the future, some algorithms that seem promising may be given too much research attention at the expense of others, and they may end up being broken or not efficient for implementation. This might happen because energy consumption is usually ignored during security standardizations, whereas security and performance are usually given top priority. To mitigate this risk, there should be close collaboration between industry, academia, and the standardization bodies to guide research and implementation directions.

10.6.3 FUTURE TRENDS AND EMERGING TECHNOLOGIES

Quantum information is gaining a lot of interest due to its secure application in the post-quantum field. The threat posed by quantum computing to the current cryptographic algorithms necessitates a shift to PQC. To elaborate, this transition is crucial to protect data from future quantum attacks. Using the harvest now, decrypt later strategy, and Shor's algorithm, traditional encryption will be easily broken by a quantum computer.

The current research is focused on various quantum-safe approaches, including quantum key distribution, hash-based signatures, and other post-quantum cryptographic algorithms. There is a significant focus on securing IoT devices and applications using PQC. One of the challenges in implementing quantum-safe protocols in IoT is the difficulty of implementing the post-quantum protocols in a constrained resource environment. To illustrate, IoT devices have a restricted amount of computation, and the quantum-safe algorithm requires a high amount of power to be performed [42]. IoT devices like sensors or smart devices have limited processing power and memory, requiring lightweight cryptography to work efficiently on these types of devices. Most of the existing system infrastructures are built around classical algorithms, so switching to PQC algorithms could lead to compatibility issues. Also, changing the infrastructure requires training the developers and updating the policies and standardization of the system.

All of these issues could impact the performance of the system. One of the recommended research focuses is on developing specialized hardware to better support PQC and optimizing the software to reduce the resource footprint. Quantum computers are still in the early stages of development is still there no clear global standardization for cryptosystems.

10.7 CONCLUSION

This chapter has provided an overview of the current state of research and development of PQC for application in embedded systems. The chapter gave an overview of what PQC is and the need for it in securing embedded devices in the foreseeable future where quantum computers have become common. We discussed the concepts behind quantum computing and PQC. The urgent need for PQC was presented, and standardization initiatives to develop PQC primitives were also discussed, including the US NIST initiative. A timeline of the NIST initiative was presented, and the current stages of the process were discussed including the selected candidates and finalists. We then provided an overview of the current state of research on implementing PQC on embedded systems, starting by discussing the threat that quantum computers pose to embedded systems, and discussing successful implementations of PQC on the embedded systems by other researchers. We finally discussed the challenges to implementing these cryptographic algorithms on embedded devices, recommended solutions and research directions, and presented future and emerging technologies in this area of research.

REFERENCES

1. Nathalie P De Leon, Kohei M Itoh, Dohun Kim, Karan K Mehta, Tracy E Northup, Hanhee Paik, BS Palmer, Nitin Samarth, Sorawis Sangtawesin, and David W Steuerman. Materials challenges and opportunities for quantum computing hardware. *Science*, 372(6539):eabb2823, 2021.

2. Christopher Monroe, Michael G Raymer, and Jacob Taylor. The US National quantum initiative: From act to action. *Science*, 364(6439):440–442, 2019.

3. Andreas Bayerstadler, Guillaume Becquin, Julia Binder, Thierry Botter, Hans Ehm, Thomas Ehmer, Marvin Erdmann, Norbert Gaus, Philipp Harbach, Maximilian Hess, et al. Industry quantum computing applications. *EPJ Quantum Technology*, 8(1):25, 2021.

4. Mohammed Saleh Ali Muthanna, Ammar Muthanna, Ahsan Rafiq, Mohammad Hammoudeh, Reem Alkanhel, Stephen Lynch, and Ahmed A. Abd El-Latif. Deep reinforcement learning based transmission policy enforcement and multi-hop routing in qos aware lora IoT networks. *Computer Communications*, 183:33–50, 2022.

5. Chafik Berdjouh, Mohammed Charaf Eddine Meftah, Abdelkader Laouid, Mohammad Hammoudeh, and Akshi Kumar. Pelican gorilla troop optimization based on deep feed forward neural network for human activity abnormality detection in smart spaces. *IEEE Internet of Things Journal*, 10(21):18495–18504, 2023.

6. Sana Belguith, Nesrine Kaaniche, and Mohammad Hammoudeh. Analysis of attribute-based cryptographic techniques and their application to protect cloud services. *Transactions on Emerging Telecommunications Technologies*, 33(3):e3667, 2022. e3667 ett.3667.

7. Oskar van Deventer, Nicolas Spethmann, Marius Loeffler, Michele Amoretti, Rob van den Brink, Natalia Bruno, Paolo Comi, Noel Farrugia, Marco Gramegna, Andreas Jenet, et al. Towards European standards for quantum technologies. *EPJ Quantum Technology*, 9(1):33, 2022.

8. Lidong Chen. Cryptography standards in quantum time: New wine in old Wineskin? *IEEE Security & Privacy*, 15(4):51, 2017.

9. David Joseph, Rafael Misoczki, Marc Manzano, Joe Tricot, Fernando Dominguez Pinuaga, Olivier Lacombe, Stefan Leichenauer, Jack Hidary, Phil Venables, and Royal Hansen. Transitioning organizations to post-quantum cryptography. *Nature*, 605(7909):237–243, 2022.

10. Manish Kumar. Post-quantum cryptography algorithm's standardization and performance analysis. *Array*, 15:100242, 2022.

11. Gorjan Alagic, Gorjan Alagic, Daniel Apon, David Cooper, Quynh Dang, Thinh Dang, John Kelsey, Jacob Lichtinger, Yi-Kai Liu, Carl Miller, et al. Status Report on the First Round of the NIST Post-Quantum Cryptography Standardization Process. US Department of Commerce, National Institute of Standards and Technology, 2022.

12. Furkan Aydin, Aydin Aysu, Mohit Tiwari, Andreas Gerstlauer, and Michael Orshansky. Horizontal side-channel vulnerabilities of post-quantum key exchange and encapsulation protocols. *ACM Transactions on Embedded Computing Systems (TECS)*, 20(6):1–22, 2021.

13. Duc-Thuan Dam, Thai-Ha Tran, Van-Phuc Hoang, Cong-Kha Pham, and Trong-Thuc Hoang. A survey of post-quantum cryptography: Start of a new race. *Cryptography*, 7(3), 2023.

14. Hamid Nejatollahi, Nikil Dutt, Sandip Ray, Francesco Regazzoni, Indranil Banerjee, and Rosario Cammarota. Post-quantum lattice-based cryptography implementations: A survey. *ACM Computing Surveys (CSUR)*, 51(6):1–41, 2019.

15. Nibedita Kundu, Kunal Dey, Pantelimon Stănică, Sumit Kumar Debnath, and Saibal Kumar Pal. Post-quantum secure identity-based encryption from multivariate public key cryptography. In *Security and Privacy: Select Proceedings of ICSP 2020*, pages 139–149. Springer, 2021.

16. Raj Badhwar. The need for post-quantum cryptography post-quantum cryptography (PQC). In *The CISO's Next Frontier: AI, Post-Quantum Cryptography and Advanced Security Paradigms*, pages 15–30. Springer, 2021.

17. Anjaneyulu Endurthi, Pallavi Yarra, Sushma Gavireddy, and Ushasri Polishetty. Closest vector problem-based proof of work mechanism for post-quantum blockchain. In *Innovations in Computer Science and Engineering: Proceedings of the Ninth ICICSE, 2021*, pages 215–220. Springer, 2022.

18. Ritik Bavdekar, Eashan Jayant Chopde, Ashutosh Bhatia, Kamlesh Tiwari, Sandeep Joshua Daniel, et al. Post quantum cryptography: Techniques, challenges, standardization, and directions for future research. *arXiv preprint arXiv:2202.02826*, 2022.

19. Mojtaba Bisheh-Niasar, Reza Azarderakhsh, and Mehran Mozaffari-Kermani. High-speed NTT-based polynomial multiplication accelerator for post-quantum cryptography. In *2021 IEEE 28th Symposium on Computer Arithmetic (ARITH)*, pages 94–101, 2021. https://doi.org/10.1109/ARITH51176.2021.00028

20. Zhaohui Chen, Emre Karabulut, Aydin Aysu, Yuan Ma, and Jiwu Jing. An efficient non-profiled side-channel attack on the crystals-dilithium post-quantum signature. In *2021 IEEE 39th International Conference on Computer Design (ICCD)*, pages 583–590. IEEE, 2021.

21. Jaime Señor, Jorge Portilla, and Gabriel Mujica. Analysis of the ntru post-quantum cryptographic scheme in constrained iot edge devices. *IEEE Internet of Things Journal*, 9(19):18778–18790, 2022.

22. Quentin Berthet, Andres Upegui, Laurent Gantel, Alexandre Duc, and Giulia Traverso. An area-efficient sphincs+ post-quantum signature coprocessor. In *2021 IEEE International Parallel and Distributed Processing Symposium Workshops (IPDPSW)*, pages 180–187. IEEE, 2021.

23. Shuzhou Sun, Rui Zhang, and Hui Ma. Efficient parallelism of post-quantum signature scheme sphincs. *IEEE Transactions on Parallel and Distributed Systems*, 31(11):2542–2555, 2020.

24. Daniel J Bernstein and Tanja Lange. Post-quantum cryptography. *Nature*, 549(7671):188–194, 2017.

25. Anshika Vaishnavi and Samaya Pillai. Cybersecurity in the quantum era-a study of perceived risks in conventional cryptography and discussion on post quantum methods. In *Journal of Physics: Conference Series*, volume 1964, page 042002. IOP Publishing, 2021.

26. George Tasopoulos, Jinhui Li, Apostolos P Fournaris, Raymond K Zhao, Amin Sakzad, and Ron Steinfeld. Performance evaluation of post-quantum tls 1.3 on resource-constrained embedded systems. In *International Conference on Information Security Practice and Experience*, pages 432–451. Springer, 2022.

27. Brian Koziel, Mehran Mozaffari Kermani, and Reza Azarderakhsh. Post-quantum cryptographic hardware and embedded systems. In *Emerging Topics in Hardware Security*, pages 229–255. Springer, 2020.

28. Manuel Barbosa, Gilles Barthe, Xiong Fan, Benjamin Grégoire, Shih-Han Hung, Jonathan Katz, Pierre-Yves Strub, Xiaodi Wu, and Li Zhou. Easypqc: Verifying post-quantum cryptography. In *Proceedings of the 2021 ACM SIGSAC Conference on Computer and Communications Security*, pages 2564–2586, 2021.

29. Tiago M Fernández-Caramés. From pre-quantum to post-quantum iot security: A survey on quantum-resistant cryptosystems for the internet of things. *IEEE Internet of Things Journal*, 7(7):6457–6480, 2019.

30. Sweta Gupta, Kamlesh Kumar Gupta, Piyush Kumar Shukla, and Mahendra Kumar Shrivas. Blockchain-based voting system powered by post-quantum cryptography (bbvsp-pqc). In *2022 Second International Conference on Power, Control and Computing Technologies (ICPC2T)*, pages 1–8. IEEE, 2022.

31. Konstantina Miteloudi, Joppe W. Bos, Olivier Bronchain, Björn Fay, and Joost Renes. PQ. V. ALU. E: Post-quantum RISC-V custom ALU extensions on Dilithium and Kyber. In *International Conference on Smart Card Research and Advanced Applications, pages 190–209*. Springer. 2023.

32. Markku-Juhani O Saarinen. Mobile energy requirements of the upcoming nist post-quantum cryptography standards. In *2020 8th IEEE International Conference on Mobile Cloud Computing, Services, and Engineering (MobileCloud)*, pages 23–30. IEEE, 2020.

33. Sean Zakrajsek. Performance analysis of nist round 2 post-quantum cryptography public-key encryption and key-establishment algorithms on ARMV8 IoT devices using supercop. 2020. https://repository.stcloudstate.edu/msia_etds/104/

34. Maximilian Schöffel, Frederik Lauer, Carl C Rheinländer, and Norbert Wehn. Secure iot in the era of quantum computers—Where are the bottlenecks? *Sensors*, 22(7):2484, 2022.

35. Lukas Malina, Lucie Popelova, Petr Dzurenda, Jan Hajny, and Zdenek Martinasek. On feasibility of post-quantum cryptography on small devices. *IFAC-PapersOnLine*, 51(6):462–467, 2018.

36. Rameez Asif. Post-quantum cryptosystems for internet-of-things: A survey on lattice-based algorithms. *IoT*, 2(1):71–91, 2021.

37. Gandeva Bayu Satrya, Yosafat Marselino Agus, and Adel Ben Mnaouer. A comparative study of post-quantum cryptographic algorithm implementations for secure and efficient energy systems monitoring. *Electronics*, 12(18):3824, 2023.

38. Kevin Bürstinghaus-Steinbach, Christoph Krauß, Ruben Niederhagen, and Michael Schneider. Post-quantum tls on embedded systems: Integrating and evaluating kyber and sphincs+ with Mbed TLS. In *Proceedings of the 15th ACM Asia Conference on Computer and Communications Security*, pages 841–852, 2020.

39. Fabio Borges, Paulo Ricardo Reis, and Diogo Pereira. A comparison of security and its performance for key agreements in post-quantum cryptography. *IEEE Access*, 8:142413–142422, 2020.

40. Dimitrios Sikeridis, Panos Kampanakis, and Michael Devetsikiotis. Assessing the overhead of post-quantum cryptography in tls 1.3 and ssh. In *Proceedings of the 16th International Conference on emerging Networking EXperiments and Technologies*, pages 149–156, 2020.

41. Qian Guo, Denis Nabokov, Alexander Nilsson, and Thomas Johansson. Sca-ldpc: A code-based framework for key-recovery side-channel attacks on post-quantum encryption schemes. *Cryptology ePrint Archive*, 2023.

42. Mohammed Mudassir, Devrim Unal, Mohammad Hammoudeh, Farag Azzedin, et al. Detection of botnet attacks against industrial iot systems by multilayer deep learning approaches. *Wireless Communications and Mobile Computing*, 2022.

11 Assessing Quantum Integer Factorization Performance with Shor's Algorithm

Junseo Lee

11.1 INTRODUCTION

Quantum information theory encompasses various topics such as quantum communication, quantum error correction, and quantum simulations. The ability to manipulate and control quantum systems has brought about the development of new technologies and the potential to solve problems that are computationally challenging for classical computers. Over the past few decades, there has been significant progress in quantum computing hardware and algorithms; however, a fully fault-tolerant quantum computer in its perfect form does not yet exist.

As computer systems evolve in both hardware and software aspects, the potential threat of quantum technologies in cybersecurity has been evident for several years. Therefore, understanding cryptographic algorithms from the perspective of quantum information theory has become a significant task in this field. It is well-known that traditional cryptographic algorithms such as the RSA cryptosystem, elliptic curve cryptography, Diffie-Hellman key exchange, and signature schemes are no longer considered secure with the emergence of integer factorization using Shor's algorithm [1] and searching using Grover's algorithm [2]. Here, we raise a practical question: How significant is the potential threat in the context of current quantum computing resources?

Given the potential of quantum information theory and the increasing development of quantum technologies, a variety of research studies on quantum information theory are presented to explore its potential effect in the cybersecurity field [3–7]. For example, research on Shor's algorithm [3, 6] explained its capability of breaking the widely used RSA public key cryptosystem developed by Rivest, Shamir, and Adleman [8], which takes advantage of the inherent potential of quantum computers. The advent of Shor's algorithm attracted significant interest from the public, resulting in a thorough analysis of the commercialization of quantum computers. This is due to the algorithm's ability to efficiently factor large numbers, potentially breaking modern classical cryptography systems. In addition, several studies [4, 7] evaluated quantum

DOI: 10.1201/9781003475286-11

attacks on the Advanced Encryption Standard (AES) [9] developed by NIST using the ability of an exhaustive key search technique by Grover's algorithm [2].

With substantial support from various literature sources, the performance of quantum computing algorithms is assessed under specific conditions. Typically, such evaluations occur in simulated quantum computing environments aimed at solving particular mathematical problems. One prominent challenge is Shor's algorithm, which is scrutinized for its potential cybersecurity implications, given its status as the most well-known method for integer factorization—an essential assumption underlying many encryption methods believed to be resistant to classical computing. However, previous studies are limited in their focus on solving specific values through tailored optimization approaches [10]. While case-by-case experiments shed light on the potential threat landscape, fully comprehending the scale of this threat using current quantum computing resources remains challenging from a broader perspective.

This study systematically evaluates the performance of integer factorization using Shor's algorithm in a gate-based quantum computing environment. To accomplish this, we conducted tests using the IBM quantum simulator, employing preselection of random parameters when performing integer factorization with Shor's algorithm. This pre-selection allows us to conduct scalable evaluations of Shor's algorithm across various ranges of N (odd square-free semiprime numbers).

11.2 THEORETICAL BACKGROUND

This section provides the required background knowledge on Shor's algorithm, *Quantum Fourier Transform (QFT)*, and matrix product state quantum simulation.

11.2.1 SHOR'S ALGORITHM

Shor's algorithm is a well-known example of a quantum algorithm for factoring integers. The detailed steps of Shor's algorithm are as follows:

1. Choose a random integer a between 1 and $N - 1$.
2. Compute the Greatest Common Divisor (GCD) of a and N using the Euclidean algorithm.
3. If GCD is not equal to 1, then we have found a nontrivial factor of N and we are done.
4. If GCD is equal to 1, use the quantum period-finding subroutine to find the period r of the function $f(x) = a^x \bmod N$. (Please refer to Section 11.3.2 for further details.)
5. If r is odd, go back to step 1 and choose a different a.
6. If $a^{r/2} = -1 \bmod N$, go back to step 1 and choose a different a.
7. Otherwise, calculate the GCD of $a^{r/2} \pm 1$ and N. If either GCD is a nontrivial factor of N, we have successfully factored N.

The most important step among them is the calculation of the smallest r, where we need to find the period r (s.t. $f(r) = 1$) of the function $f(x) = a^x \bmod N$ (step 4).

Table 11.1

The Types of IBM Simulators and the Theoretically Breakable Bits β Depend on Their Supported Number of Qubits Q

Quantum Simulator (Type)	Q	β
Statevector simulator (Schrödinger wavefunction)	32	7
Stabilizer simulator (Clifford)	5000	1249
Extended stabilizer simulator (Extended Clifford (e.g., Clifford+T))	63	15
MPS simulator (Matrix Product State)	100	24
QASM simulator (General, context-aware)	32	7

To achieve this, we use a quantum period-finding subroutine. The quantum period-finding algorithm employs the QFT to efficiently find the period r of the function $f(x)$. The QFT is the quantum analog of the classical Fourier transform and is a key component of various quantum algorithms. It enables the transformation of a quantum state encoding a function into a superposition of its periodic components.

11.2.2 QUANTUM FOURIER TRANSFORM

The QFT is a quantum algorithm that performs a Fourier transform on a quantum state. It is used in various quantum algorithms, including Shor's algorithm for factoring integers and Grover's algorithm for searching an unsorted database. The QFT is similar to the classical Fourier transform, but it uses quantum gates and operations to perform the transformation. The QFT takes a quantum state $|x\rangle$, where x is a binary string of length n, and maps it to another quantum state $|y\rangle$, where y is a binary string of length n. The transformation is defined as follows:

$$\text{QFT}|x\rangle = \frac{1}{\sqrt{2^n}} \sum_{y=0}^{2^n-1} e^{\frac{2\pi i x y}{2^n}} |y\rangle \tag{11.1}$$

The QFT can be implemented using a series of quantum gates, including the Hadamard gate, phase shift gates, and controlled-phase shift gates.

11.2.3 MATRIX PRODUCT STATE (MPS) QUANTUM SIMULATION

The IBM quantum circuit simulators are software tools that enable users to simulate the behavior of quantum computers on classical computers. These simulators are meticulously crafted to deliver precise and efficient simulations of quantum circuits, empowering researchers and developers to thoroughly test and optimize their algorithms prior to deploying them on actual quantum hardware. Table 11.1 presents an overview of the currently supported IBM simulators, specifying the maximum number of qubits they can accommodate, as well as the theoretically breakable bits in the RSA scheme using Shor's algorithm.

This study specifically utilized the MPS simulator for simulating quantum circuits. With support for up to 100 qubits, it facilitated the evaluation of Shor's algorithm for integer factorization of numbers up to 24 bits. In comparison, the stabilizer simulator, which supports the most qubits, was limited to simulating Clifford gates and lacked support for non-Clifford gates essential for effectively implementing Shor's algorithm.

In particular, MPS simulations are known to be advantageous for quantum circuits with weak entanglement [11]. The fundamental element of the MPS simulation protocol entails locally decomposing the general pure state of the n spins, denoted as $|\Psi\rangle \in \mathscr{H}_2^{\otimes n}$, into n tensors $\{\Gamma^{[l]}\}_{l=1}^n$ and $n-1$ vectors $\{\lambda^{[l]}\}_{l=1}^{n-1}$ within the given two-dimensional Hilbert space $\mathscr{H}_2$. Here, the tensor $\Gamma^{[l]}$ is assigned to qubit l and possesses at most three indices, denoted as $\Gamma_{\alpha\alpha'}^{[l]i}$, where α, α' range from 1 to χ (here, χ is termed the *Schmidt number*, signifying the maximum Schmidt rank across all conceivable bipartite partitions of the n qubit system), and i ranges from 0 to 1. Conversely, $\lambda^{[l]}$ represents a vector with its elements $\lambda_{\alpha'}^{[l]}$ storing the Schmidt coefficients of the partitioning $[1:l]:[(l+1):n]$. Accordingly, the quantum state of multiple particles across N sites is represented by the following structure:

$$|\Psi\rangle = \sum_{i_1=0}^1 \cdots \sum_{i_n=0}^1 c_{i_1 \cdots i_n} |i_1\rangle \otimes \cdots \otimes |i_n\rangle, \qquad (11.2)$$

where the coefficients $c_{i_1 \cdots i_n}$ are computed as:

$$c_{i_1 \cdots i_n} = \sum_{\alpha_1 \cdots \alpha_{n-1}} \Gamma_{\alpha_1}^{[1]i_1} \lambda_{\alpha_1}^{[1]} \Gamma_{\alpha_1 \alpha_2}^{[2]i_2} \lambda_{\alpha_2}^{[2]} \cdots \Gamma_{\alpha_{n-1}}^{[n]i_n}, \qquad (11.3)$$

with the constraint $\sum_{i_1=0}^1 \cdots \sum_{i_n=0}^1 |c_{i_1 \cdots i_n}|^2 = 1$.

The memory requirements for the computation increase proportionally to $\chi^2 n$, while efficient simulation of quantum circuits in terms of memory and computational demands is feasible when $\chi \sim \text{poly}(n)$, signifying weak entanglement. Accordingly, MPS simulations can vary in computational and memory requirements depending on the Schmidt number between two qubits, and the structure of quantum circuits such as the order of quantum registers can significantly impact simulation time. Further analysis on this matter will be conducted in Section 11.3.3.

11.3   INTEGER FACTORIZATION USING SELECTED PARAMETER

11.3.1   THE EFFECT OF RANDOM PARAMETERS

Random parameters play a crucial role in the effectiveness of Shor's algorithm. Specifically, the selection of a random parameter a in the first step is pivotal for the algorithm's efficiency. If a random number sharing a common factor with N is chosen, the Greatest Common Divisor (GCD) calculated in the second step will exceed 1, enabling the direct identification of a nontrivial factor of N. However, opting for a random number coprime to N results in a GCD of 1, necessitating reliance on the quantum period-finding subroutine to determine the period r of the function $f(x) = a^x \bmod N$. Hence, the algorithm's success hinges on the stochastic selection of parameters and the probabilistic nature of quantum measurement.

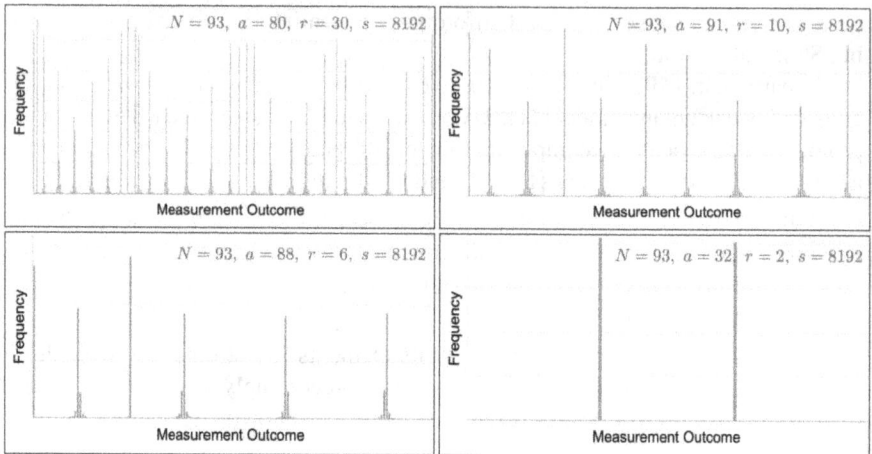

Figure 11.1   Simulation results of the quantum period-finding subroutine in Shor's algorithm when varying N, a, r.

The outcomes of the quantum period-finding subroutine in Shor's algorithm are depicted as probability distributions based on measurement outcomes, as illustrated in Figure 11.1. In an ideal quantum computing environment with a sufficient number of shots, the number of histogram bars would precisely correspond to the value of r. However, in the presence of noise, as observed in the figures, errors are introduced when employing a noisy quantum simulator or quantum hardware. Insufficient iterations can render the interpretation of experimental findings challenging. It is anticipated that larger values of r will necessitate more iterations. For instance, when $r = 2$, only two bars appear in the histogram, allowing for the attainment of the quantum period-finding subroutine's outcome with a limited number of iterations. Figure 11.1 exhibits the results obtained from factoring $N = 93$, with corresponding values of $a = 80, 91, 88$, and 32, leading to respective r values of 30, 10, 6, and 2. Here, s denotes the number of shots for the quantum circuit.

Thus, we pre-select the value of a with a minimum value of r (which is 2) and use it instead of randomly selecting a for scalable testing of Shor's algorithm. However, this method also has limitations as N increases. Therefore, we want to compare the time required to factor using a quantum simulator as N increases while restricting a to the same value of r. Table 11.2 shows several values of a that make r equal to 2 for actual semiprime N. To measure the performance of the simulator under standardized conditions, the (N, a) pairs listed below are executed together instead of selecting a random a.

Table 11.2

Examples Values of a Where $r = 2$

N	15	129	335	687	7617	9997
a	4	44	66	230	2540	768

11.3.2 QUANTUM ORDER FINDING ROUTINE

The quantum order finding routine in Shor's algorithm is implemented based on the following steps:

1. *Prepare the quantum state*: You first need to prepare a quantum state that is a superposition of all possible values of the period of the function $f(x) = a^x \bmod N$. This involves applying a Hadamard gate to a set of qubits that will represent the period.

2. *Apply the modular exponentiation function*: Apply the modular exponentiation function to the state, which maps the state $|x\rangle$ to $|a^x \bmod N\rangle$. This involves applying controlled gates that implement the function $f(x) = a^x \bmod N$. Specifically, for each qubit in the period register, you apply a controlled-$U_{a^{2^j}} \bmod N$ gate to the state, where U_a is a unitary that performs the function $f(x) = a^x \bmod N$.

3. *Apply the inverse QFT*: Apply the inverse QFT to the period register, which maps the state $|r\rangle$ to a state that is proportional to $|\frac{r}{f}\rangle$, where r is the period and f is the order of the Fourier transform. This involves applying a set of controlled-phase gates that depend on the position of the qubits in the period register.

4. *Measure the period register*: Measure the period register and use classical post-processing to find the period. Specifically, you measure the period register and obtain a value r.

5. *Continued fractions algorithm*: Take the ratio $\frac{r}{f}$ and create a continued fraction $[a_0, a_1, a_2, a_3, \cdots]$. Determine the whole number component of the continued fraction as $a_0 = \lfloor \frac{r}{f} \rfloor$. Adjust the fraction to be $\frac{r}{f} = \left(\frac{r}{f} - a_0 \right)^{-1}$. Determine the next whole number component a_i as $a_i = \lfloor \left(\frac{r}{f} - a_{i-1} \right)^{-1} \rfloor$. Adjust the fraction to be $\left(\frac{r}{f} - a_{i-1} \right)^{-1} = a_i + \left(\frac{r}{f} - a_i \right)^{-1}$. Continue doing this until the fraction becomes a repeating sequence or until a maximum number of iterations is reached. The period of the continued fraction is the length of the repeating sequence. If the length of the repeating sequence is odd, repeat the last value in the sequence to make it even. The period of the continued fraction is the smallest value of r that satisfies the equation $a^r \bmod N = 1$.

11.3.3 ENTANGLEMENT ANALYSIS FOR MPS

As discussed in Section 11.2.3, the arrangement order of quantum registers is crucial for efficient simulation. In circuits implementing Shor's algorithm, registers are
broadly categorized into upper, lower, and ancilla. Therefore, by systematically
changing their order and computing the von Neumann entropy, one can assess the
degree of entanglement between one register and the rest. Figure 11.2 illustrates the
experimental results corresponding to the case of $N = 15$ and $a = 4$. Analyzing these
experimental results reveals that the entanglement between the Upper register and
the combined Lower+Ancilla registers is consistently strong on average. Hence, for
MPS simulation, it is appropriate to consider the order of Upper–Lower–Ancilla
or Upper–Ancilla–Lower.

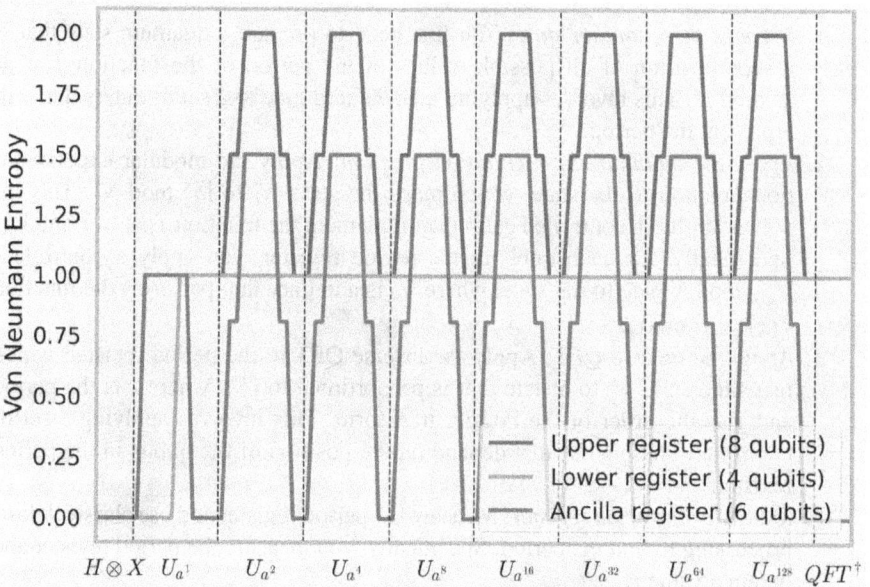

Figure 11.2   Analysis of the degree of entanglement for each register.

11.4 PERFORMANCE OF SHOR'S ALGORITHM AT SCALE

In this section, the time taken for integer factorization using Shor's algorithm on the
IBM quantum circuit simulator is presented under two scenarios: pre-selection of
parameter a and random selection of parameter a. All factorization times represent
the duration required for the aforementioned quantum period-finding subroutine.

    Figure 11.3 illustrates the performance of Shor's algorithm in terms of the re-
quired time for integer factorization given various inputs of N. The integer factorization time when parameter a is pre-selected is depicted linearly, showing an overall
tendency on a log-log scale. While Shor's algorithm theoretically exhibits polyno-

mial time complexity with respect to $\log N$, the results obtained through classical simulation deviate from this prediction. The pattern remains generally consistent within the same number of bits, with a significant increase in the required time observed as the number of bits increases.

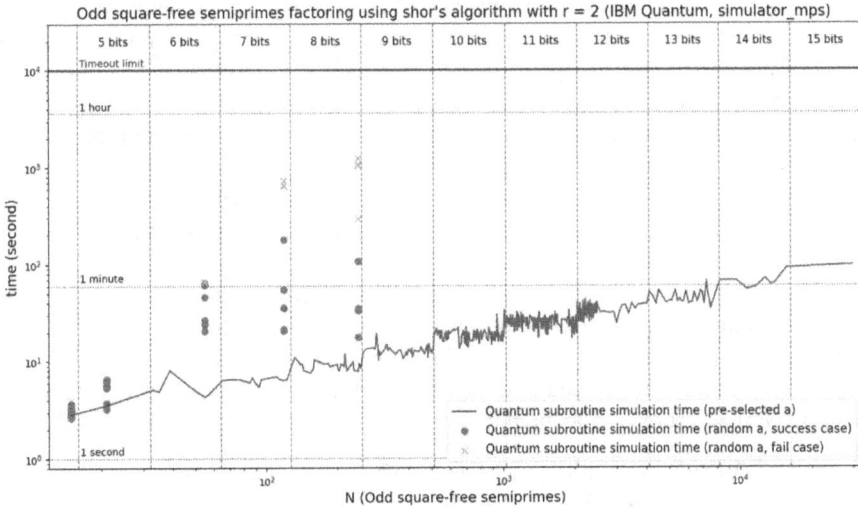

Figure 11.3   The result of the quantum period-finding subroutine time

One noteworthy aspect of the graph is the scalability of performance measurement for integer factorization with respect to N. This implies that the measurements were not limited to specific numbers but were applicable to all possible values of N, enabling the prediction of the time required for integer factorization for any given N. This is highly significant as it allows for the prediction of the feasibility and time requirements of integer factorization for all possible values of N.

Furthermore, it is worth noting that all experimental results were obtained with a fixed number of eight shots. When a is pre-selected with a minimum period of $r = 2$, eight shots were sufficient to factorize all given input numbers.

On the other hand, in the case of random selection of a, the value of r is not consistent, making it impossible to measure the factorization time consistently. The red circles on the graph demonstrate significant variations in the factorization time for the same number, depending on the trial, and the yellow X marks indicate instances of failure to produce a factorization result. (In this experiment, the timeout limit was set to 10,000 seconds.) Within this timeout limit, when pre-selection is performed, successful factorization was achieved up to 14 bits. However, when random selection was used, it was observed that successful factorization could not be achieved for numbers with 9 bits or more.

11.5 CONCLUSION AND FUTURE WORKS

In this study, we evaluated the scalability of integer factorization performance using Shor's algorithm in quantum circuit simulation based on the MPS method. Comparing the performance of integer factorization between pre-selection and random selection of parameters, we concluded that pre-selecting a is beneficial for scaling quantum computing performance. Furthermore, it is crucial to be mindful of the impact of parameter a on the performance of Shor's algorithm, as it determines the period of the function to be evaluated. This variable is expected to affect both the time required to generate quantum circuits and the simulation (computation) time of quantum circuits. Further research is needed to analyze the optimal parameter selection methodology through time complexity analysis for each step (quantum circuit generation, quantum circuit operation, and continued fraction processing) in order to identify the most effective parameters.

Moreover, researchers have utilized the ε-random technique to optimize the number of quantum samples required for algorithms in Quantum Random Access Memory (QRAM) [12] based on Lemma 1. Building on this technique, we intend to carry out theoretical research aimed to reduce the complexity of integer factorization and explore various factorization methodologies beyond Shor's algorithm.

Lemma 1 (Lévy's inequality [13]) *Let $f : S^n \to \mathbb{R}$ be a function defined on the n-dimensional hypersphere S^n with Lipschitz constant*

$$\eta = \sup_{\mathbf{x}_1, \mathbf{x}_2} \frac{|f(\mathbf{x}_1) - f(\mathbf{x}_2)|}{\|\mathbf{x}_1 - \mathbf{x}_2\|} < \infty, \tag{11.4}$$

with respect to the Euclidean norm $\| \cdot \|$ and a point $\mathbf{x} \in S^n$ be chosen uniformly at random. Then,

$$\Pr\left[|f(\mathbf{x}) - \mathbb{E}[f]| \geq \varepsilon\right] \leq 2\exp\left(-\frac{C(n+1)\varepsilon^2}{\eta^2}\right), \tag{11.5}$$

for some constant $C > 0$.

REFERENCES

1. Shor, P. W. (1994, November). Algorithms for quantum computation: discrete logarithms and factoring. In Proceedings 35th Annual Symposium on Foundations of Computer Science (pp. 124–134). IEEE.

2. Grover, L. K. (1996). A fast quantum mechanical algorithm for database search. In Proceedings of the Twenty-Eighth Annual ACM Symposium on Theory of Computing (pp. 212–219). https://arxiv.org/abs/quant-ph/9605043

3. Shor, P. W. (1999). Polynomial-time algorithms for prime factorization and discrete logarithms on a quantum computer. SIAM Review, 41(2), 303–332.

4. Bonnetain, X., Naya-Plasencia, M., & Schrottenloher, A. (2019). Quantum security analysis of AES. IACR Transactions on Symmetric Cryptology, 2019(2), 55–93.

5. Jordan, S. P., & Liu, Y. K. (2018). Quantum cryptanalysis: shor, grover, and beyond. IEEE Security & Privacy, 16(5), 14–21.

6. Gerjuoy, E. (2005). Shor's factoring algorithm and modern cryptography. An illustration of the capabilities inherent in quantum computers. American Journal of Physics, 73(6), 521–540.

7. Grassl, M., Langenberg, B., Roetteler, M., & Steinwandt, R. (2016). Applying Grover's algorithm to AES: quantum resource estimates. In International Workshop on Post-Quantum Cryptography (pp. 29–43). Cham: Springer International Publishing.

8. Rivest, R. L., Shamir, A., & Adleman, L. (1978). A method for obtaining digital signatures and public-key cryptosystems. Communications of the ACM, 21(2), 120–126.

9. Dworkin, M. J., Barker, E. B., Nechvatal, J. R., Foti, J., Bassham, L. E., Roback, E., & Dray Jr, J. F. (2001). Advanced encryption standard (AES). https://nvlpubs.nist.gov/nistpubs/FIPS/NIST.FIPS.197.pdf

10. Jiang, S., Britt, K. A., McCaskey, A. J., Humble, T. S., & Kais, S. (2018). Quantum annealing for prime factorization. Scientific Reports, 8(1), 17667.

11. Vidal, G. (2003). Efficient classical simulation of slightly entangled quantum computations. Physical Review Letters, 91(14), 147902.

12. Jeong, K. (2023). Sample-size-reduction of quantum states for the noisy linear problem. Annals of Physics, 449, 169215.

13. Lévy, P., & Pellegrino, F. (1951). Problémes concrets d'analyse fonctionnelle (2nd edition). Gauthier-Villars.

12 Quantum Computing Based Attacks on Cryptography and Countermeasures

Devrim UNAL, Abdulrahman AlRaimi,
Sandrik Concepcion Das, and Saad Mohammed Anis

12.1 INTRODUCTION

Quantum computing is known to negatively affect cyber security by weakening the security of many cryptographic algorithms. However, it also introduces a world with exciting rules that could be exploited to our benefit in this post-quantum era and with quantum cryptography [1]. We could say that the largest impact of quantum mechanics is, exerting a pressure in the realm of cyber security. Because of this, we need to prepare for the post-quantum era by developing new protocols and encryption schemes that are resilient to quantum and classical computers. Quantum computing is the reason behind this research move, which will lead to more robust and better cyber security in the future. A good example is the B92 protocol, which will be explained in this chapter.

12.1.1 B92 PROTOCOL

The first Quantum Key Distribution Protocol that saw the light is BB84 protocol by Charles Bennett and Gilles Brassard in 1984. B92 protocol has been developed based on BB84 protocol by Charles Bennett in 1992. The aim for B92 protocol is to make it easier and less complex than BB84 protocol [3]. To start explaining the B92 protocol, we need to know some basics. Simply, the detector of the photon has two bases, $+$ *and* $\times$. If the detector is adjusted to detect a photon on the $+$ base, it will detect either a vertically polarized light or a horizontally polarized light. If the detector is adjusted to detect a photon on the $\times$ base, it will detect a diagonally polarized light, either with $+45°$ or $-45°$. To start this protocol, Alice will use either a horizontally polarized photon ($\rightarrow$) which will represent the 0 value, or a diagonally polarized photon with $+45°$ ($\nearrow$) which will represent the 1 value. In the other hand,

Table 12.1

Combinations of Polarized Light and Detector Bases for the B92 protocol

Polarized Photon	Detector Base	Detected Photon	Value
→	+	→	...
→	×	↗	...
→	×	↘	0
↗	+	↑	1
↗	+	→	...
↗	×	↗	...

Bob will use either $+$ or $\times$ base to detect the coming photon. But how will Bob determine the value of the coming photon [17]?

12.1.2   POLARIZED LIGHT AND DETECTORS

From Table 12.1, we understand the following:

- If a photon is detected using the $+$ base, there are two options:
  - If → is detected, then either → or ↗ has been sent. So, we cannot determine the value.
  - If ↑ is detected, then 100% ↗ has been sent. So, the value is 1.
- If a photon is detected using the × base, there are two options:
  - If ↗ is detected, then either → or ↗ has been sent. So, we cannot determine the value.
  - If ↘ is detected, then 100% → has been sent. So, the value is 0.

So, a transmission of a secret key will be like this:

- For each photon, Alice Polarize it horizontally or diagonally with $+45°$.
- Alice sends all photons to Bob.
- Bob detects each photon with $+$ or $\times$ base.
  - If the detected photon is ↑, then the value is 1.
  - If the detected photon is ↘, then the value is 0.
  - Otherwise, no value is assigned.
- Bob tells Alice the result of his measurements.

Alice and Bob retain the corrected measured photons and discards the rest.

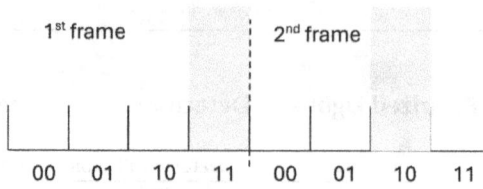

Figure 12.1   A modulated frame using 4-PPM technology

12.1.3   IMPROVED B92 PROTOCOL WITH PULSE POSITION MODULATION (PPM)

It has been proved that the effective transmission for B92 Protocol is 25%, meaning that 75% of the measurement is discarded, which means that the key rate is very low. Pulse position modulation improves the key rate [3]. We will not discuss this technology here, but we will discuss its effect. Using PPM, we could transform each bit to represent n bits. The idea is very simple. We will choose first how many bits we want each optical pulse to represent. Say we chose 2. Then, it was proved that we need at least $2*n$ time slots to represent n bits. In our case, we need four time slots. The idea is to have a frame and divide it into $2n$ time slots, therefore, four time slots in our case. Then encode the optical pulse in one of these time slots. Each time slot represents a value that could be represented using two bits. Imagine you are a receiver, for each frame, you will have four possible values. If the optical pulse is encoded in:

> 1st time slot: the value is: $0 \rightarrow 00$
> 2nd time slot: the value is: $1 \rightarrow 01$
> 3rd time slot: the value is: $2 \rightarrow 10$
> 4th time slot: the value is: $3 \rightarrow 11$

A transmission of multiple photons in the B92 Protocol with the 4-PPM will be like this (See Figure 12.1):

1. For each optical pulse, Alice prepares a frame with two time slots.
2. Alice polarizes each optical pulse horizontally or diagonally with $+45^o$.
3. Each polarized optical pulse will be encoded on a random time slot on its associated frame.
4. Alice sends all frames to Bob.
5. Bob detects each pulse with $+$ or $\times$ base.

   a. If the detected pulse is ↑ or ↘, then get the value using 4-PPM rules.
   b. Otherwise, no value is assigned.

6. Bob tells Alice the result of his measurements.
7. Alice and Bob retain the corrected measured photons and discards the rest.

12.1.4 SECURITY OF THE B92 PROTOCOL

To transfer a light particle between Alice and Bob, we need a channel. A lossy channel is an ideal channel where there is no noise. Noise in a channel is something that causes a bit error. An example of bit error is when Alice sends 1 while Bob detects 0 or vice versa. It was proven that B92 protocol is unconditionally secure, and any attempts of eavesdropping are detected in a lossy channel [5]. However, we know that in real life, nothing is perfect. So, we need to analyze the security of B92 in a noisy channel.

12.1.5 COLLECTIVE-ROTATION NOISE CHANNEL

One type of noisy channel is the Collective-Rotation noise channel. A particle transmitting through this channel will deflect with angle θ. One important thing to note is no direct way to distinguish between eavesdropping error and noise error. So, in order to detect eavesdropping, we need to determine the initial qubit error rate ε. The initial qubit error is the maximum noise rate that could happen to a particle in the absence of eavesdropping. If noise rate has been computed for a certain communication and it was higher than ε, then an eavesdropping act has occurred 100% [4]. Before diving into the result of this analysis, we introduce these variables:

- ber_0 is the initial qubit error ε.
- ber_1 is the bit error rate when the eavesdropper (Eve) detects a photon successfully, either $\searrow$ or $\uparrow$, and resend it directly to Bob.
- ber_2 is the bit error rate when the eavesdropper (Eve) detects a photon successfully, either $\searrow$ or $\uparrow$, and send either $\nearrow$ or $\longrightarrow$ to Bob.

Since detailed proofs are behind the scope of this chapter, we will introduce the results directly. After the analysis, we observe the following:

- $ber_0 = sin^2(\theta) = \varepsilon$
- $ber_1 = \frac{7+8\varepsilon(1-\varepsilon)}{16}$
- $ber_2 = \frac{1}{4}+\frac{1}{2}\varepsilon$

From Figure 12.2, there are two cases:

- If $0.5 \leq \varepsilon \leq 0.56$, then $ber_1 \geq ber_0 \geq ber_2$
- If $\varepsilon \leq 0.5$, then $ber_1 \geq ber_2 \geq ber_0$

From point 1, we understand that we could detect the eavesdropping act only if The eavesdropper, Eve, resends what she detected directly. If Eve chooses always to send either $\searrow$ or $\uparrow$, there is no way to detect it. Therefore, a channel with an initial qubit error greater than 0.5 must be avoided because the noise rate is very high, and we cannot detect Eve. However, we understand from point 2 that a channel with an initial qubit error smaller than 0.5 is secure because any eavesdropping act will increase the noise rate higher than the initial qubit error, so Eve will always be detected. Assume an eavesdropping act existed on a channel with an $\varepsilon \leq 0.5$; the

Noise Analysis

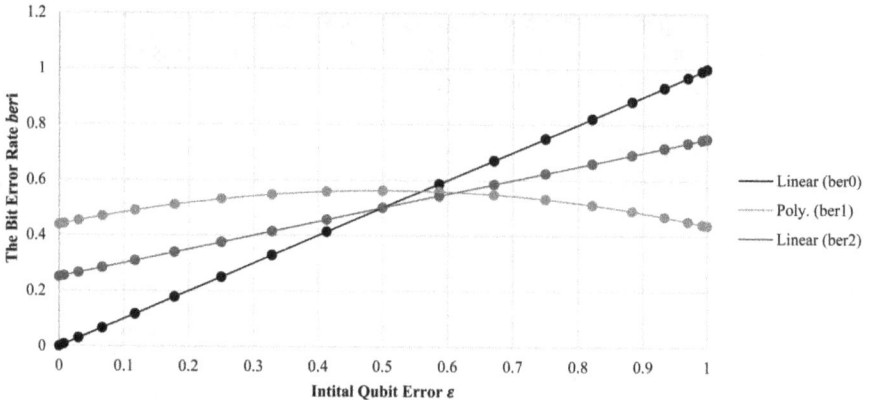

Figure 12.2   Collective Rotation Noise Channel

most amount of information Eve can gain will be 50%, and she will not be able to know what she gets. Even if she knows what she gets, she only got 50% of the key. Hence, the B92 protocol is secure when $\varepsilon \leq 0.5$.

12.2   NEGATIVE IMPACT OF QUANTUM COMPUTING ON CRYPTOGRAPHY

While quantum computing shows high potential toward the advancements in computational processing and simulation, the same power it possesses also renders dangerous shortcomings.

12.2.1   ISSUES WITH QUANTUM PARTICLES

One of the issues that arise from the properties of quantum bits, also known as "qubits" is called quantum decoherence. Decoherence is the loss of the ability of qubits to remain in quantum superposition and/or entanglement. Qubits are very sensitive to their environment and any slight disturbances could result in decoherent qubits, which essential lose their ability to be in a bi-state superposition. Hence, quantum computers are specifically built in void conditions such as sub-zero temperatures and vacuumed spaces.[6]

12.2.2   VULNERABILITY OF EXISTING CRYPTOGRAPHIC TECHNIQUES

Asymmetric encryption: Asymmetric public key encryption methods such as the Rivest, Shamir, and Adleman (RSA) and the Diffie-Hellman (DH), which rely on the principles of discrete logarithms and large integer factorization being hard to decipher, can be easily dissolved on quantum computers. A quantum technique recognized as Shor's algorithm can quickly render these algorithms vulnerable through the process. Shor's algorithm works around the principle that qubits can hold a superposition of all possible values (in this case, prime factors), thus exposing the two private numbers following specific computational steps.

Symmetric encryption: With respect to symmetric encryption, which uses the same private key and cipher to encrypt and decrypt information (bidirectional), an algorithm developed by Lov Grover can be implemented on a quantum computer to find a specific target element within the order of $O(\sqrt{N})$ operation, compared to a conventional classical computer which could complete the search in $O(N)$ steps. For example, working with 128-bit key length would be around the security level of a 64-bit key length cipher. However, this vulnerability can be slightly mitigated by increasing the key sizes by two times minimum.

Hash functions: Generally, a hash function convert a large random set of inputs into a fixed set output and is known to operate in a unidirectional manner. Since the produced output is of a fixed size, Grover's algorithm can detect a collision in $O(\sqrt{N})$ steps, similar to popular symmetric ciphers. Nonetheless, hashes such as SHA-2 and SHA-3 stand resistant to quantum techniques.[2]

12.3   GROVER'S ALGORITHM

Symmetric encryption is an encryption scheme where Alice encrypts a message with a key, and Bob decrypts this message with the same key. One way to find the key is the exhaustive search which could be achieved by trying all possible keys. For example, a message p has been encrypted with a key k of 16-bits which produced a ciphertext c. In exhaustive search, we need to try $2^{16}-1$ keys in the worst case. So, the time complexity for exhaustive search is $O(2^n)$ where n is the key size. In today's world, Advanced Encryption System (AES) is integrated into most applications as the main way to encrypt messages. In AES, a message could be encrypted with either 128-bit, 192-bit, and 256-bit key sizes. Without a doubt, there is no classical computer in the world that could find a key for a message that has been encrypted using AES-128 in a reasonable amount of time. The best-known attack on full AES is the key recovery attack which needs to try 2^{126} keys for AES-128, which is still infeasible. However, this is only the case with a classical computer. In Quantum world, Grover algorithm can find the key with just $O(\sqrt{2^n})$. So, instead of trying 2^{128} possible keys, we will try 2^{64} possible keys. To understand how huge this improvement is, think about the ratio between these two numbers:

$$2^{64}.2^{128}$$

$$1 :\sim 18 \times 10^{18}$$

Therefore, for each 1 key that will be tested in Quantum Computer, 18×10^{18} keys will be tested in classical computer. Which means that Grover algorithm is an enormous success and a dangerous threat to the classical symmetric encryption schemes in general.

States

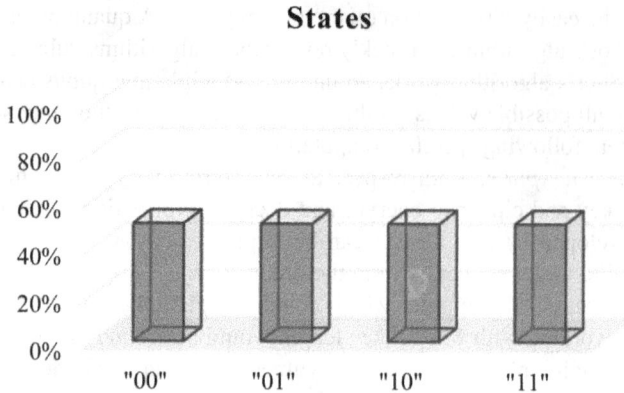

100%
80%
60%
40%
20%
0%

"00"   "01"   "10"   "11"

Figure 12.3 Initialization Step

12.3.1   ORACLE FUNCTION IN GROVER'S ALGORITHM

Grover algorithm is a general quantum search algorithm for an unstructured database. The core for Grover algorithm is the Oracle function. To understand what Oracle function does, imagine we have a list from 1 to 4. In this example, we need to find the number 3. The Oracle function is a function that produces 1 when we put a solution and 0 otherwise. So,

$$f(x) = 0, x \neq 3 \text{ and } f(3) = 1$$

Generally, the oracle function is defined as follows:

$$f(x) = 1, \text{ when } x = a \ (a \ solution)$$

$$f(x) = 0, \text{ otherwise } (no \ solution)$$

For every certain problem, we need to create an Oracle function that follows the criteria above and integrate it to a Grover algorithm [15]. The following are the steps of Grover algorithm; we have list from 1 to 4, and we need to find number 3.

12.3.2   GROVER ALGORITHM STEPS

Firstly, how may qubits we need? Notice that we can represent each number in our list in terms of 2 bits, 00, 01, 10, 11. So, we need in total 2 qubits [14].

Initialization Step: As seen in Figure 12.3, we need first to initialize the 2 Qubit to zero then we put each Qubit in the superposition state. By doing this, we are giving each possible solution the same probability to be chosen. For example, the probability of choosing 00 is 25% just as the probability of choosing 01, 10, or 11.

States

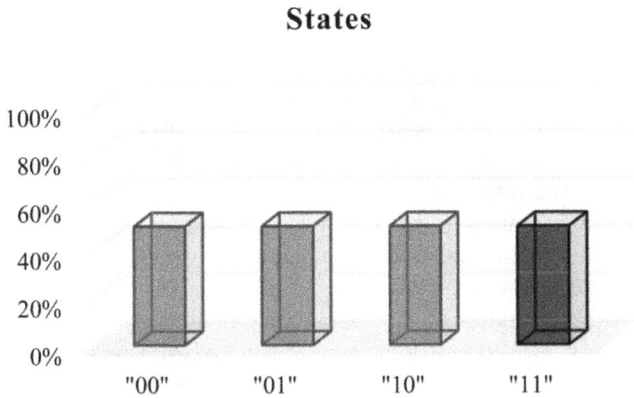

Figure 12.4 Oracle Function Step

States

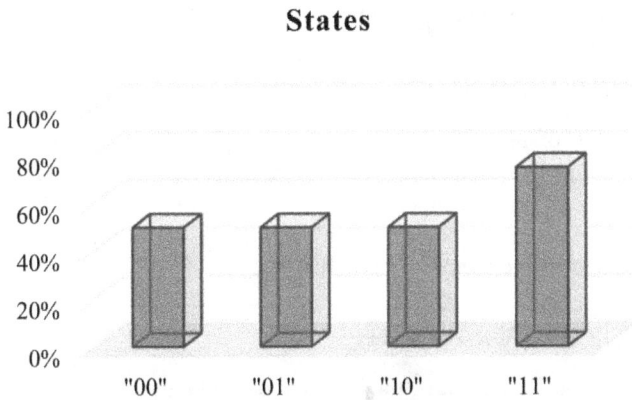

Figure 12.5 Diffusion Operator Step

Oracle Function Step: All the qubits, which are in superposition state, will enter the oracle function. As seen in Figure 12.4, the oracle function will simply "mark" the most possible solution. In our case the solution is 11.

Diffusion Operator Step: Since we have a state that could be a solution, now we need to give more probability of success (see Figure 12.5). To do this, we will use something called Grover diffusion operator, which does this exactly.

Repetition: We will simple repeat steps 2 and step 3 $\sqrt{2^n}$ times. In our case, we will repeat the steps $\sqrt{2^2} = 2$.

Measurement: This is the final step. In this step, we will measure the result. In our case, the solution will be 11 with a probability that is near 100%.

12.3.3   GROVER ALGORITHM ON SIMPLIFIED AES

It has been shown that Grover algorithm could be used to find a key for a plaintext that have been encrypted using a simplified version of AES called SAES. In SAES, the key size is 16 bits. One might say, it is only 16 bits, where is the risk? The answer is that it could theoretically enlarge to any number of bits with only one limitation being hardware. The main reason why researchers underestimate the value of the Grover algorithm is that you need either a Universal Quantum Computer (UQC) or AES Quantum Circuit integrated into a Quantum computer for that specific purpose. However, UQC do not exist, but a AES Quantum Circuit has been built. The purpose of an AES Quantum Circuit is that an adversary can integrate it later as an Oracle function. Considering that we need to build a SAES with the lowest number of Qubits, we needed around 64 qubits. Results have been shown that an adversary will find the key for a certain text that has been encrypted with AES after completing $\frac{\pi}{4}\sqrt{2^n}$ iterations only. Therefore, theoretically, AES is in danger. However, in real life, we are still in the early stages of Quantum Computing. Meaning, we might see the benefit of the Grover algorithm, or any similar algorithm after a decade or so. However, we need to prepare for its effects from now [7].

12.4   SHOR'S ALGORITHM

The implications of using quantum computing, particularly in cryptography, are speeding up the process not through lower resource demand but rather by reducing the number of steps during computation. As mentioned in earlier sections, the heart of asymmetric encryption methods, such as the RSA, is based on the significant resource demand to factorize large integers.

12.4.1   SHOR'S ALGORITHM STEPS:

To better understand the process, the classical procedure of integer factorization using Shor's algorithm is explained in the following subsection [8]. Let us assume that N, the number we are trying to factorize, has only two factors, p and q.

Step 1: Start with a guess value, a, such that $a < N$, and a and N are co-prime, i.e., they do not share any common factors with each other. If $gcd\,(a,N) \neq 1$, therefore, a is a factor of N, and we can perform normal division, $N \div a$, to find the other factor, hence solving the factorization.

Step 2: Compute the period of the function,

$$f_{(a,N)}\,(x) = a^x \bmod N.$$

Call this period r. If r is odd, redo step 1.

Step 3: Since the period r is assumed to be even, check:

$$a^{\frac{r}{2}} + 1 \neq 0 \bmod N.$$

Step 4: We know that $a^r = 1 \ mod \ N$. Therefore, $a^r - 1 = 0 \ mod \ N$. Also, For some given multiple k of N, $a^r - 1 = k \times N$. Also, since r is even, therefore:

$$\left(a^{\frac{r}{2}} - 1\right)\left(a^{\frac{r}{2}} + 1\right) = k \times N = k \cdot (p \cdot q).$$

We can then assume that:
$p = gcd(a^{\frac{r}{2}} - 1, N)$ and $q = gcd(a^{\frac{r}{2}} + 1, N)$.

12.4.2 QUANTUM IMPLEMENTATION

The quantum implementation of Shor's algorithm mainly requires two co-entangled quantum registers, R1 and R2. Register R1 will roughly contain an adequate number of qubits needed to represent the integers from 0 until N-1 in a superposition or until a noticeable pattern can be observed to find the period. The function $f_{(a,N)}(x) = a^x \ mod \ N$ will be evaluated for all values of x till N, and the results will be stored in the register R2. [13] The rest of the implementation is fundamentally identical to the classical procedure since the main point of the quantum implementation is to calculate the period of the stated modulo function in a shorter number of operations. The implementation also involves techniques, such as applying the Hadamard gate, black-box operation, and quantum Fourier transformation onto register R1 [9].

12.5 FUTURE RESEARCH TRACK

With regards to the coming decade, compact quantum computers seem to be out of reach due to hardware and environmental limitations. Since qubits have short coherent times, therefore the short effective usage time to work with them drastically hinders processing complex computations. Another field that is open to further research is error correction or fault-tolerant quantum systems. This sector holds significant priority for development since qubits are hypersensitive to their environments and measurement of quantum readings are done through unorthodox protocols. Moreover, simply increasing the number of qubits does not contribute toward better results without lowering their error probabilities.

Despite all these challenges, enterprises like IBM and Google have ambitious plans to upscale quantum computers. As of 2024, IBM has made significant advancements in quantum computing. The company successfully released a 1,000-qubit quantum chip in late 2023 and is now focusing on scaling up further. IBM's roadmap includes plans to build a quantum computer with over 4,000 qubits by 2025, leveraging a modular architecture in their next-generation Quantum System Two, which is designed to integrate multiple processors and enable quantum-centric supercomputing. Such a landmark would be able to mitigate the fundamental issues in the current rudimentary state of quantum technology and would pave the path toward quantum supremacy [10]. Researchers have also theorized fault-rectifying conventions that involve multiple Qubit encoding from a single qubit [11].

Another department that could use further research is cryogenic semiconductor technology. Since some quantum architectures rely on Quchips (short for "quantum chips") consisting of qubits, control circuits, and measuring devices, it is most likely that they are in a cryogenic setting altogether. This brings the need for cryogenic semiconductor technologies that fulfill the required functionalities that can withstand low temperatures without inducing internal system noise and producing erroneous qubits [12].

Most of the above-discussed security mechanisms based on quantum computing are resource and energy intensive [16], therefore they cannot be applied onto IoT systems due to resource and computational limitations [17]. IoT systems have high mobility which require specific computational models for modeling security aspects [18]. Especially, security policy models need to be designed and implemented according to the requirements of IoT devices, to provide secure devices without reducing its computational performance [19]. There is not much current work in the area of security modeling and security policy verification for quantum computing on mobile systems.

12.6 CONCLUSION

Quantum computing holds the promise to revolutionize the state of modern computing and encryption. However, quantum technology still requires decades of evolution and concentration to be integrated into daily life. The theory is developing steadily as multiple algorithms have been designed to solve particular problems more efficiently than classical computers. These algorithms have built the base on which further research could expand and meet future demands. Hardware developments, however, did not catch up with the speed of the theoretical work. The existence of a universal quantum computer is still an exceptionally far-off target. One of the significant obstacles in quantum computing is noise in quantum channels – a practical issue that ideal quantum computers do not possess. The developed algorithms assumed to be working on optimal machines strike an imbalance with the noisy practical examples. The power of quantum computing can help the world to improve computation astronomically, but it also holds a considerable risk of damage to the modern world. This technology, as researchers, should be handled, discovered, and exploited carefully to reap its eventual reward. With the technology steadily blooming, it is a great time to delve in as we expand our horizons into unparalleled computing power.

REFERENCES

1. Hidary, J. D. (2019). Quantum Computing: An Applied Approach (1st ed., 2019 ed.). Springer.

2. Mavroeidis, V., Vishi, K., Zych, M. D., & Jøsang, A. (2018). The Impact of Quantum Computing on Present Cryptography. International Journal of Advanced Computer Science and Applications, 9(3), 405–414. https://doi.org/10.14569/ijacsa.2018.090354

3. Xiong, Y., & Lai, H. (2019). The Improved Protocol Based on the Bennett 92 Protocol. ACM TURC '19: Proceedings of the ACM Turing Celebration Conference, 1–8. https://dl.acm.org/doi/10.1145/3321408.3326689

4. Li, L., Li, J., Li, C., Li, H., Yang, Y., & Chen, X. (2019). The Security Analysis of Quantum B92 Protocol in Collective-Rotation Noise Channel. International Journal of Theoretical Physics, 58(4), 1326–1336. https://doi.org/10.1007/s10773-019-04025-7

5. Ali, N., Mat Radzi, N. A. N., Aljunid, S. A., & Endut, R. (2020). Security of B92 Protocol with Uninformative States in Asymptotic Limit with Composable Security. AIP Conference Proceedings 2203, 4–5. https://doi.org/10.1063/1.5142141

6. Herman, A., & Friedson, I. (2018). Quantum Computing: How to Address the National Security Risk (1st ed.) [E-book]. Hudson Institute. http://media.hudson.org.s3.amazonaws.com/files/publications/Quantum18FINAL3.pdf

7. Almazrooie, M., Abdullah, R., Samsudin, A., & Mutter, K. N. (2018). Quantum Grover Attack on the Simplified-AES. ICSCA 2018: Proceedings of the 2018 7th International Conference on Software and Computer Applications, 204–211. https://doi.org/10.1145/3185089.3185122

8. Ardhyamath, P., Naghabhushana, N. M., & Ujjinimatad, R. (2019). Quantum Factorization of Integers 21 and 91 Using Shor's Algorithm. World Scientific News, 123, 102–113. http://psjd.icm.edu.pl/psjd/element/bwmeta1.element.psjd-160c7d48-5b72-448d-8959-39fdeb6cf551

9. Veliche, A. (2018). Shor's Algorithm and Its Impact On Present-Day Cryptography.

10. Cho, A. (2020). IBM Promises 1000-Qubit Quantum Computer—A Milestone-by 2023. Science, 15. https://www.sciencemag.org/news/2020/09/ibm-promises-1000-qubit-quantum-computer-milestone-2023 (Retrieved Sep. 15, 2020).

11. Cho, A. (2020). The Biggest Flipping Challenge in Quantum Computing. Science. https://www.sciencemag.org/news/2020/07/biggest-flipping-challenge-quantumcomputing (Retrieved Aug. 3, 2020).

12. Chang, C. R., Lin, Y. C., Chiu, K. L., & Huang, T. W. (2020). The Second Quantum Revolution with Quantum Computers. AAPPS Bulletin, 30(1), 9–22. http://aappsbulletin.org/myboard/read.php?Board=featurearticles&id=236

13. Quantum Factoring Algorithm. (2020, May 9). [Video]. YouTube. https://www.youtube.com/watch?v=9p9jW9ee4JE (Retrieved Apr 8, 2021).

14. QuTech. (n.d.). Code Example: Grover's Algorithm. Quantum Inspire. https://www.quantum-inspire.com/kbase/grover-algorithm/ (Retrieved April 8, 2021).

15. Grover's algorithm. (n.d.). Qiskit. https://qiskit.org/textbook/ch-algorithms/grover.html (Retrieved Apr 8, 2021).

16. QKD (B92 protocol). (n.d.). QuVis, University of St. Andrews. https://www.st-andrews.ac.uk/physics/quvis/simulations_html5/sims/cryptography-b92/B92_photons.html (Retrieved Apr 8, 2021)

17. Zubair, M., Unal, D., Al-Ali, A., & Shikfa, A. (2019). Exploiting Bluetooth Vulnerabilities in E-Health IoT Devices. In *Proceedings of the 3rd International Conference on Future Networks and Distributed Systems*. Association for Computing Machinery. https://doi.org/10.1145/3341325.3342000

18. Unal, D., & Caglayan, M. U. (2006). Theorem Proving for Modeling and Conflict Checking of Authorization Policies. In *2006 International Symposium on Computer Networks, Istanbul*, pp. 146-151. https://doi.org/10.1109/ISCN.2006.1662524

19. Unal, D., & Caglayan, M. U. (2013). Spatio-Temporal Model Checking of Location and Mobility Related Security Policy Specifications. Turkish Journal of Electrical Engineering and Computer Sciences, 21, 144–173. https://doi.org/10.3906/elk-1105-54

Index

For Product Safety Concerns and Information please contact our EU
representative GPSR@taylorandfrancis.com
Taylor & Francis Verlag GmbH, Kaufingerstraße 24, 80331 München, Germany

www.ingramcontent.com/pod-product-compliance
Lightning Source LLC
Chambersburg PA
CBHW070713220326
41598CB00024BA/3133